The Manager's Environment

An Economic Approach

G. H. Webster B Sc (Econ) attended the London School of Economics where he specialized in industry and trade. He worked in industry for eight years and is now Senior Lecturer in Economics in the School of Management, Ealing Technical College. J. M. Oliver, B Sc (Econ), M Phil, read economics at University College, London, and has taught in secondary and further education. He is now Principal Lecturer in Economics in the School of Business and Social Studies, Ealing Technical College. They are also the authors of *Public Policy and Economic Theory*.

Editorial Advisory Board

M. A. Pocock, B Sc (Econ) *Chairman*;
E. P. Gostling; H. Shearing;
A. H. Taylor, MC.

Also available in this series

MARKETING MANAGEMENT IN ACTION – *Victor P. Buell*
THE AGE OF DISCONTINUITY – *Peter F. Drucker*
THE EFFECTIVE EXECUTIVE – *Peter F. Drucker*
MANAGING FOR RESULTS – *Peter F. Drucker*
THE PRACTICE OF MANAGEMENT – *Peter F. Drucker*
CYBERNETICS IN MANAGEMENT – *F. H. George*
PLANNED MARKETING – *Ralph Glasser*
FINANCE AND ACCOUNTS FOR MANAGERS – *Desmond Goch*
HOW TO WIN CUSTOMERS – *Heinz M. Goldmann*
BUSINESS PLANNING – *D. R. C. Halford*
INNOVATION IN MARKETING – *Theodore Levitt*
MAKING MANPOWER EFFECTIVE (Part 1) – *James J. Lynch*
CAREERS IN MARKETING – *Institute of Marketing*
MARKETING – *Colin McIver*
SELLING AND SALESMANSHIP – *R. G. Magnus-Hannaford*
THE PROPERTY BOOM – *Oliver Marriott*
THE MANAGER AND THE ORGANIZATION – *Eric Moonman*
EXPORTING – *Robin Neillands and Henry Deschampneufs*
DYNAMIC BUSINESS MANAGEMENT – *Harold Norcross*
FINANCIAL PLANNING AND CONTROL –
R. E. Palmer and A. H. Taylor
GUIDE TO SAMPLING – *Morris James Slonim*
COMPUTERS FOR MANAGEMENT – *Peter C. Sanderson*
MY YEARS WITH GENERAL MOTORS – *Alfred P. Sloan, Jr*
THE REALITY OF MANAGEMENT – *Rosemary Stewart*
MANAGERS AND THEIR JOBS – *Rosemary Stewart*
MANAGEMENT DECISION-MAKING – *Edited by Gordon Yewdall*
MANAGEMENT INFORMATION –
Gordon Yewdall, G. P. Mead, C. T. Wicks and C. W. Smith

CONDITIONS OF SALE

This book shall not, by way of trade or otherwise, be lent, re-sold, hired out or otherwise circulated without the publisher's prior consent in any form of binding or cover other than that in which it is published and without a similar condition including this condition being imposed on the subsequent purchaser. The book is published at a net price, and is supplied subject to the Publishers Association Standard Conditions of Sale registered under the Restrictive Trade Practices Act, 1956.

Management Series

The Manager's Environment

An Economic Approach

G. H. WEBSTER
and
J. M. OLIVER

A PAN ORIGINAL

PAN BOOKS LTD · LONDON

First published 1970 by Pan Books Ltd,
33 Tothill Street, London, S.W.1

ISBN 0 330 02638 0

© G. H. Webster and J. M. Oliver 1970

Printed in Great Britain by
Richard Clay (The Chaucer Press), Ltd, Bungay, Suffolk

Contents

To our parents

Preface

This book is for managers and is written with special regard to non-economists. It has its origins in our experience of teaching short management courses but should be of interest to a wider group.

Picture a short course on general management waiting to start a session on 'The Economic Environment'. What could be said which would be relevant? The authors have considerable experience of this situation, from the stage, as it were, rather than from the stalls, and have asked themselves the same question. They have sometimes felt themselves in a variety show, rather than the theatre, in which the time allowed has been too short for them to show the full extent of their act and their position in the bill hints that the main event occurs later. This is not to belittle the role of such a session in a course of this type – the audience has usually been appreciative – still less to claim that it should have 'star billing'. It serves to identify both the readers and the authors of this book.

Why should a manager bother to read a book of this type? Firstly, he may be interested for private, personal reasons, in the sense that any increase in his knowledge may give him satisfaction and anyway he exists in this environment in two separate roles – that of employee and private citizen. Secondly, since, without some knowledge of the wider environment, his firm may apparently be subjected to random and inexplicable pressures. For both these reasons, a better understanding of his environment is likely to result in a better manager, one who appreciates the present situation and can therefore weigh the alternatives more accurately.

What we have learnt from courses is the degree of ignorance of many managers about the economic environment in which they operate, coupled with considerable interest in learning more. This is not meant to be patronizing: like all teachers we have learnt a good deal from our students. What we, as authors, have tried to do is to explain the economic environment to a manager who has never studied economics – *from the point of view of the economist.* A glance at the Contents Page will show that we have tried to do more than write a description; we have also been concerned with how and why the economy works. Partly this reflects a difference between a book and a face-to-face session. Despite our, sometimes desperate, attempts to learn something about the industrial background against which managers on a particular specialist course operate, we can never match their local knowledge. Therefore the value of our teaching depends heavily on audience participation to show the immediate relevance of the more general points which we can make. Obviously in a book this two-way communication cannot operate; but the problem is eased in two ways. Firstly, we have not tried to describe particular industries or markets. Secondly, we have, on the whole, been encouraged by the ease with which course members have been able to apply our generalities to their specific cases.

Some words of warning: however hard we have tried to be scientific and objective, in some ways this book repeats some of our social (rather than political) prejudices. It also reflects our interests, although the interest of course members in the past has been a much more important criterion of what to include, what to leave out, and at what depth to deal with a particular topic.

Sheer pressure on space has meant that some topics have had to be excluded; for example, apart from a brief look at the significance of the Balance of Payments, international affairs have been ignored. Our determination not to recount recent economic history has meant that some elements in the economy, usually government policies which change very quickly, have been ignored or looked at largely in principle. An example of the first is Prices and Incomes Policy, and, of the second, credit control and 'squeezes'.

Some features of a book on the economic environment are best explained right away. It will contain facts, theories to explain why things happen, and policies to meet certain goals.

The facts we hope are accurate; a point that has come from sessions which we have run is just how many simple statements about the economy are unknown or are obtainable only in a distorted form. We must also admit that there are many simple factual questions about the economic environment to which no answer can be given. All the data presented in the book are the most recent available to us; older figures which have had to be used in places make the point that we often have too little information, available too late. The theories outlined will, we hope, always be relevant, although necessarily heavily simplified and will reflect the current mainstream thought of economists.

The goals and policies are likely to be more familiar to the reader; although perhaps we will harp on the difference between the two (ends and means) rather more than is often the case. In this area we are on more contentious ground since we are dealing with 'value judgements' (jargon for opinions) rather than positive statements about scientific relationships. The economist describing the environment can do more than discover facts and give his views on them; he has some theories which are as well proved as many in the physical sciences. Perhaps the reader's view of the book will depend on *his* value judgement regarding the relevance of the explanations behind the facts.

Finally, it is with considerable pleasure that we are able to mention the help which we have received in writing this book – apart from authors of quotations who are individually acknowledged at the end of the chapters. Our thanks go first to Mr M. A. Pocock for his encouragement and helpful suggestions which have gone beyond his responsibilities as Editor of this series. Many of our friends have read parts of drafts but we would like to thank in particular our colleagues Mrs J. Fawcett and Messrs G. Janata, J. H. Lloyd, F. J. Rendall, M. F. J. Simpson, A. Smith and A. H. Taylor. We believe that their advice has greatly improved the quality of our performance; any errors and shortcomings of presentation continue to be our joint responsibility. We are most appreciative of the friendly service provided by Mrs I. Hirst and the staff of our college library. Our typists, Mrs U. Walker and Mrs J. Browne, have coped admirably with the problems which they have had to face. We are very grateful for all the help which we have received.

G. H. WEBSTER
J. M. OLIVER
Ealing Technical College, 1970

CHAPTER ONE

Objectives and Policies

Introduction

As the preface has stated this book is written by economists for non-economists. But the book is not written just for any non-economist; it is specifically designed for those in management positions. The treatment is non-technical in the sense that it assumes no prior knowledge by the reader.

What can the economist offer the manager?

Firstly, the economist has at his disposal vocational expertise in such fields as marketing and operational research – this is not our present concern.

Secondly, the economist looks at the environment in a special and helpful kind of way. This is far from a unique claim; so do historians, townplanners, and many others. It is not our position that the economist's viewpoint is better than that of other professions – it is our position that the economist's understanding can be an advantageous part of the businessman's equipment. A manager's work is improved by a deeper understanding of his economic environment; this is not to say that all economists make good managers but that a lack of an economist's understanding will limit the skills of, and policies open to, the manager – to his own disadvantage and that of his firm. This is not then a book on management economics but on economics for managers; there is not much concern with technical problems within the firm but rather with the environment of the firm.

This is not an economics textbook in the conventional sense of producing economists; all it might do is give the reader a look at the world through economist's eyes. The aspects of the environment which economists can clarify for managers are easily detailed as, briefly and non-technically, the pattern of resource use, the level of economic activity, and economic growth – and the way in which these three aspects of the environment modify the strategies of the firm.

Let us take each in turn. An *individual*'s level of *economic activity* is a familiar enough concept which manifests itself, for instance, in whether he is unemployed, whether he works part time, full time, or whether overtime is available. A man's income is *not* a measure of his economic activity compared to other men, it is a rough and ready measure of its value.

If, instead of comparing one man with another, we compare a given man's income over time, then higher wages do mean greater economic activity, provided his hourly wage, or piece-work rate, is constant.

The level of economic activity of a *firm* shows itself, for example, as a one-shift system where three are possible, as unused factory and warehouse space, excess stockholding, and so on. A firm which engages in labour hoarding is showing an awareness of activity levels and of expectations about them.

A low level of economic activity is indicated for the *economy* by the presence of generally high unemployment, and a high level of activity by generally rising prices. Particular prices may, of course, rise during unemployment just as particular individuals may be unemployed during inflation. In trying to give an idea of the activity level we have indicated tell-tale signs rather than presented strict definitions.

The *pattern of resource use* is a less obvious idea although it is a very general problem. If an *individual* has to choose between alternatives due to limited spending power, then he has a resource allocation problem whether it be between two normal shirts or one expensive shirt, between a holiday or buying furniture – or any other such choice. These are relatively simple choices; difficulties arise when time is introduced into the problem, as in a choice between a cheap second-hand car with high running costs and an expensive new car with low running costs. In such cases, low present and high future costs must be compared with high present and low future costs; this is not a simple matter of arithmetic, for the future costs, as we shall see, must be discounted in order to express them in present values. Allocation problems occur in a *firm* in allocating workers between production tracks, floor-space between stockholding and production, and so on. Similar problems exist within an *economy* in such choices as nuclear or oil-fired power stations and between house and hospital building.

Economic growth exists for the *individual* in rising real purchasing power with given commitments, for a *firm* in assets rising faster than the general price level and for the *economy* in rising real output per head. In all such cases, economic growth expands the range of opportunities available.

Such problems are part of a firm's environment but these examples specify the content of economic study rather than define it fully. An astronomer would not like to be summed up as 'just studying stars' since this would leave him with the astrologers. So we must say a little more to define the role of the economist than simply list his problems. It will help to define his method as that of the social scientist. For our purposes, this means two things: that the economist tries to *test* his theories about activity, allocation, and growth problems and that he distinguishes between *positive* and *normative* statements.

This is a vital dividing line; a positive statement can be verified by an appeal to facts while a normative statement cannot be verified and so is a belief, a matter of opinion or propaganda – which is not to say that it is wrong. The price paid for a Beatles record is a positive fact but its value is a matter of opinion. This distinction helps us to see the economist's role and to cut him down to size. It is a matter of opinion whether we prefer a bungalow to a house, and we would expect an architect not to impose his taste but to proffer technical advice on, for instance, which foundations are consistent with one's choice. Just so an economist cannot, as such, tell a firm whether it should set itself a goal of short-run or long-run profits but he can say which pricing policy suits which goal. The economist's role is technical in nature; since his opinions have no special virtue he is concerned with means rather than ends in the sense that he has professional expertise on means but is a layman like anybody else when it comes to ends. Once the goals are chosen – and the choice is normative – then the economist can specify policies consistent with those goals and we can ask that the specifications be positively based and adequately tested.

Having seen what it is that economists try to explain, let us see how some understanding of these matters might lead to a modification of a firm's policies.

The Firm

A firm has no intrinsic nature; it means different things to different people and takes on aspects, interests, and problems according to viewpoint. A lawyer might be interested in whether it is a private or public limited liability company or a partnership, and would not consider the Church of England to be a firm; a behavioural scientist might be interested in societal problems and use exactly the same concepts to analyse both firms and the Church of England. This latter approach is nearer that of the economist to whom a 'firm' is a collection of resources to be allocated between competing uses. He is not primarily interested in the legal or social aspects of the firm save that they define some of the boundaries to the firm's policies. For example, a choice between a labour intensive method (many office clerks) and a capital intensive method (office mechanization), available in theory, may not be available in practice to a partnership due to its restricted access to capital. Similarly, out of several production methods technically possible, some may be eliminated as unworkable for trade union reasons. The economist has no technical skill to bring to such problems but they help to define and narrow his own role. Once a firm is seen as a collection of allocatable resources, not only is the term narrowed to eliminate legal and other aspects but it is also widened to include a cricket club and the Church of England – which undoubtedly have to allocate scarce land, labour, and capital between alternative uses.

Let us look, in some detail, at a firm's allocation problems and then at those of the activity level and of growth. The economist is interested in two separate aspects of allocation problems. Firstly, in the criteria upon which these allocations are made and secondly, the criteria upon which these apportionments may be judged, as for, or against, the public interest. As we have indicated, incorporated one-man businesses, partnerships, limited liability companies, nationalized corporations, the English soccer team, and the Baptist Church are all grist to the economist's mill. These bodies can all benefit in a quite clear way from the economist's approach which has a common vision on how a car manufacturing company might allocate its design staff into different teams, one for the development of a sports car and one for saloon cars, and also on arranging a cricket team of ten field players into five or six batsmen and five or four bowlers. The next problem of how many and what sort of bowlers and whether to compromise with an all-rounder is also open to the same analysis.

Objectives of the Firm

Before we can set out the main criteria for allocation *it is necessary to specify the goal of a firm since this will itself influence the allocation.* This is immediately obvious for when a cricket team wishes to maximize its chance of not losing rather than winning it will field more seam bowlers than leg-spinners, unlike one which wishes to maximize its chance of winning at a cost of greater vulnerability. Similarly a football team's aim helps it to choose between field allocations such as 4-2-4, 4-3-3, and so on. So in allocating resources we must know our objectives. Whatever sport's journalists might say there is no 'correct' football or cricket strategy; there is only a correct policy for a particular purpose.

There is an immediate difficulty in trying to identify the

goals of a firm. Some firms clearly try to maximize profits, but this is more ambiguous than might appear at first sight, for to maximize short-run profits may not be the same as maximizing long-run profits; other possible goals (developed in Chapter Eight) are sales maximization, cash-flow maximization, or maximization of market share. Not only is it possible to generate a list of quite different goals but there is no reason to believe all firms are single-goal firms. Once this possibility is allowed, goals like profit maximization, subject to retaining a minimum market share, can be formulated; but it could have been maximizing market share subject to a minimum profit requirement. And so on almost endlessly. *The important point is that a slight re-formulation of the goal normally necessitates a re-allocation of resources.* In a credit squeeze, for instance, a firm might give a greater priority to cash-flow and thus be forced to cut down other activities. For example, this might mean lower staff bonuses and less staff recruitment.

A goal for the firm currently fashionable among academics is net present worth or net present value. If a firm seeks to maximize net present worth then it chooses that combination of present and future assets which have a greater present value than any other combination. Thus an oil firm might give a higher value to its existing stock of refineries and garages plus exploration activities in Alaska, which can only give extra profits in the future, than to extending its refineries and garages to give extra profits in the present. This is because the future Alaska profits can have a higher value, when discounted to present values, than the present value of extra refining and garage profits. (See later for an explanation of discounting.)

Our subject is becoming slippery and blurred for our approach is not concerned with the 'firm' in the ordinary sense and we find no single answer in looking for the 'right' policy of the firm. We will keep our subject in focus by sticking to positive economics and to seeking strategies consistent with goals rather than interesting, but necessarily nconclusive, discussions of 'correct' goals.

Resource Allocation within the Firm

By what criteria should a firm allocate resources between competing uses? Examples are engineers between different projects, bulldozers between sites, or land between car-park extensions and new buildings. The answer is straightforward: all activities should be extended to that level where their marginal returns are equal. The last thousand pounds (marginal thousand pounds) spent by a firm on producing tractors should make the same contribution to profits, or whatever is being maximized, as the last thousand spent on car production. *If the marginal contributions are unequal then it is possible to re-allocate from the less profitable activity to the more profitable and raise profits without changing costs and to continue this approach until marginal returns are equal.*

This principle is very general as we can see by indulging briefly in two frivolous examples. It may seem an attractive idea to field an attacking soccer formation that will score many goals, but this is simple-minded for the purpose of the game is to maximize a favourable goal ratio, not simply goals scored. There is no point in scoring five goals and conceding six. Seen in this way it is clear that a manager should field a formation whereby the last attacker scores as many goals as are saved by the last defender.

By way of example, we postulate a profit-maximizing firm allocating £10,000 in five separate sums of £1,000 to two alternatives, A and B, which generate the following profit contributions:

A:	£5,000	B:	£6,000	
	£4,000		£5,500	
	£3,000		£4,500	
	£2,000		£4,000	
	£1,500		£3,500	
TOTALS	£15,500		£23,500	*Total* of A and B £39,000

Now the marginal contributions are unequal at £1,500 and £3,500 in A and B respectively; £2,000 might be withdrawn from A and allocated to B yielding the following:

A:		B:	
	£5,000		£6,000
	£4,000		£5,500
	£3,000		£4,500
			£4,000
	£12,000		£3,500
			£3,300
			£3,000
		TOTAL	£29,800

Total of A and B £41,800

Total costs are unchanged at £10,000 but profits have risen by £2,800 due to re-allocating expenditure between A and B in the ratio 7:3, instead of 5:5, and the marginal returns are now equal in A and B at £3,000.

Similarly the last batsman in a cricket team should score as many runs as the last bowler prevents. The notions of consistency between policies and goals and of the incremental approach are two of the economist's main analytical tricks. It is a fruitless common practice to discuss policies independently of goals, a policy is neither 'right' nor 'wrong' in itself but is appropriate only to particular goals. In Chapter Six we will analyse SET in this manner.

We will now analyse such allocations using as our criteria the interests of the firm. The equi-marginal returns rule is a most useful guide and many firms try something along these general lines, and yet, without question, firms allocate resources in a sub-optimal manner from their own point of view. There are five possible reasons.

Firstly, of course, many of a firm's activities are incapable of incremental adjustment because they involve *indivisibilities*, that is inputs which come in relatively fixed sizes – it is not sensible to talk of 1½ turbines. It is worth noting that some escape can be gained by hiring, for example, a computer for one week a year rather than buying 1/52 of a computer or buying a computer and using it 1/52 of a year.

Secondly, the economist is concerned with *true costs* rather than money costs. This means seeking the alternatives forgone by a particular course of action, for the true cost to a firm of a new office block is, for example, the lorry fleet forgone. *This real cost, sometimes called opportunity cost, is seldom accurately indicated by money prices.* To take just one example, a firm in deciding whether to make, or buy, a component should not simply compare money costs of, say, 15*s* in production with a purchase price of £1 if the component is bought from outside. This approach favours internal production. It is possible that the resources allocated to production could have made something valued at 30*s* if used elsewhere and it is this forgone production which is the real cost. This more careful approach favours external purchase in our example.

A third difficulty can be that a production process can generate *externalities* which make it difficult to gauge correctly the true cost of production. The sort of thing we have in mind is the way in which the installation of new equipment on a production track can inconvenience production on a nearby track, in such ways as holding up the delivery of materials and components and so raising marginal costs. This kind of effect is very difficult to predict and to evaluate.

A fourth difficulty in measuring costs, and for that matter revenues, is created by the notion of interest rates and *time. This puts us in a position where we have to modify the values of costs and revenues according to the time at which they occur.* Thus a cost of £100 this year should not be favourably compared with £106 revenues next year if current interest rates are 7 per cent. This is because £100 of costs today equals £107 in a year's time so that, *measured at the same moment of time*, the costs exceed the revenue. It is vital to compare costs and revenues at constant prices if a sensible decision is to be made. This is normally achieved by the discounting of future sums of money back to present values. Revenues and costs cannot be properly judged without knowing the time at which they occur. This is why a currently fashionable goal of the firm is to maximize net *present* worth.

A fifth difficulty is in the assessment of *benefits*. There are, firstly, technical difficulties in the measurement of profits; such as the depreciation policy to be followed, provision for doubtful debts, and so on. But even if the 'facts' of profits are reasonably agreed we are not necessarily much further forward. The firm can seldom be sure that it has maximized profits, since it cannot be sure what revenue (and costs) might have been generated by a different pricing policy. Even if, for some reason such as legally enforced prices, no other pricing policy was available, it cannot be sure what it has done to future profits, nor that a different marketing policy, such as advertising, discounts, or after-sales service, might not have produced greater sales and profits at the given price.

It is equally difficult for a firm to judge its performance if it is trying to maximize tax losses or market share rather than profits. The problem gets out of hand altogether if we introduce time, for if the firm is concerned with long-run profits maximization it can never be quite certain whether today's profits reflect prices that will retain or offend customers, or costs that reflect production conditions that will retain or offend staff. Once we look forward from today's profits, we leave the world of facts for that of inference and probability. Economists do have theories to cope, at least provisionally, with these problems of marketing but they are outside our present concern.

So far we have tried to show three separate things. Firstly, sense can only be made of an allocation problem if the goal (maximand) can be specifically identified – for the economy it might be full employment and for the firm it might be long-run profits. Secondly, these objectives are all normative, no goal, or set of goals, can be shown to be logically superior to others: they can only be preferred as a matter of opinion. Thirdly, the achievement of a given goal is often necessarily fulfilled within limits (constraints). For the economy it might be full employment provided inflation does not exceed a particular rate. For the firm such constraints may be subjective, such as the desire to maximize profits as long as a wide product range is also offered to give greater security to a company. There are also resource constraints, production may be limited at least in the short-run by the availability of components, or staff, financial constraints such as departmental or company budgets and legal constraints such as Factory Acts or trade union agreements which limit the ways in which labour can be used.

The National Economy and its Goals

If we continue with this mode of thinking the rest quickly falls into place. The economy is a collection of resources to be allocated between competing claims in order to achieve largely inconsistent goals. The authors once asked four separate groups of business executives to specify the goals of government economic policy; no one group's list resembled that of any other and, once a consolidated list had been jointly agreed, no group could put the list in the same order of priority as any other group. The surprising thing was that they were unable to see that they could hardly expect optimal government economic policy in these circumstances and had little right to moral indignation with government policies. None of them would have dreamt of asking a design engineer to produce a car without saying whether it was to be a sports car, saloon car, or station wagon. They would have given short shrift to any executive who did so; yet they were quite happy to demand an economic policy without saying what it was for.

What are the goals of UK government economic policy? They appear to be[1] full employment, a stable exchange rate, a balance of payments surplus, price stability, economic growth, redistribution of income, and aid to underdeveloped countries. This is an inconsistent and formidable list; no country, for instance, has yet produced long-run full employment and price stability. It is hardly surprising that a society which demands a multi-goal economic policy is dissatisfied with the results. But it is anomalous that few see that they are the equivalent to a customer demanding a cheap, luxuriously equipped, fast, low petrol consuming sports car that is also an estate car.

If we set ourselves a number of goals we should not be surprised if none are fully achieved. The economist can bring special knowledge to the problem of getting a policy consistent with a goal, or a mix of goals, and by pointing out the inconsistencies and contradictions he may help to make the normative choice of goals a knowledgeable one. But such a choice can never be positive, scientific or demonstrably 'right'; it is always a matter of opinion.

The sequence of events in some company takeovers draws the attention of businessmen to the large element of estimate in profit figures, both of past and future performances, even when they are not so reminded by economists; what is less obvious is that the goals of government policy, so often taken as self-evident, are also opinions open to discussion. We will consider just five of them to illustrate not that economists know better goals but that they know something of how these goals might be evaluated.

Economic growth is valued for two main reasons. Firstly, it has become a matter of international prestige; newspapers are for ever producing league tables of international comparative performances, often with an air of nervous self-flagellation. But, usually, much less attention is paid to comparisons of living standards which are more favourable to the UK; nor to the fact that the current UK growth rate is much faster than it was for most of the Victorian era – often imagined as a vanished golden age. To seek faster growth for prestige reasons is nothing to do with economics as such. Secondly, growth is valued because it appears to lead to increased living standards and, indeed, the two terms appear synonymous. In fact, examination of the disadvantages of economic growth show this to be partly illusory. Higher living standards for whom? There is some evidence that faster growth can lead to a more unequal income distribution. This has certainly been the case in Western Germany. Economic growth means the expansion of some industries and contraction of others, the car industry expands at the expense of the bicycle industry. Some of the newly unemployed will never be employed again and others only at lower living standards. Economic growth has another cost in that it means consumption must be forgone now in favour of investment to produce greater consumption in the future. Not all those who bear the present costs will enjoy the future benefits. Lastly, there is the problem of externalities and and deterioration of amenities; if there are more supersonic aircraft and transistor radios, then more people have to listen to them and more power stations means less scenic beauty.

Economic growth, then, is not an unalloyed gain.[2]

Full employment can be more succinctly assessed. It avoids the twin disadvantages of unemployment, which are demoralizing social distress and lost production opportunities. But full employment in its turn has the disadvantages of being generally associated with inflation and of hindering the changes necessary for economic growth. It is difficult to be serious simultaneously about achieving both full employment and maximizing growth. The community needs to choose some level of employment together with some rate of growth – and also, as we shall see in a later chapter, some particular rate of inflation.

Price stability is a policy goal for four reasons. Firstly, it is held that inflation diminishes the level of savings. This is just not true; the level of income is the main determinant of savings and there are plenty of examples of savings rising during inflation. Secondly, it is held that inflation erodes the value of money savings and raises the value of other assets such as property and some company shares. This is broadly true; but there is nothing in economics to say that those who hold their wealth in money are more worthy than those who hold it in other forms. Thirdly, if one country has greater inflationary pressures than others then it has difficulties with its Balance of Payments. But this, in its turn, is a goal of policy open to discussion. So these three reasons for desiring price stability are all normative rather than positive, matters of opinion rather than matters of fact. A fourth point is that inflation bears hard on those on fixed incomes such as pensioners. This is quite true and has to be weighed against the disadvantages of other goals. In any event, what would be said of a system whereby the greater growth that can accompany inflation led to higher pensions? The problems of fixed income groups are not unchanged facts of life, nor confined to inflationary periods; they are, or can be, the subject of social policy, to attend to them means to forgo something else, but in the end, they exist because we choose to let them exist.

It is more tortuous to discuss stability of the exchange rate, that is to say avoiding devaluation as a goal of government policy – devaluation is an especially emotive topic to many people. It must first be made clear that many other countries take a quite different view; Bolivia, for instance, has devalued many times in the post-1945 era and such patterns are quite common in less-developed countries which use devaluation as a policy technique rather than seek to avoid it.

The non-economist should remember three points about devaluation. Firstly, currency-rate devaluation is often confused in popular discussion with domestic price rises and both are described in emotive phrases like 'the value of sterling'. It may be that avoiding both inflation and devaluation are widely supported and legitimate goals of economic policy, but they are quite separate, both technically and conceptually, and they need different evaluations and appraisals; nothing whatsoever is gained by pretending they are the same thing. Secondly, a currency devaluation is a price change like any other price change but it is a price change between countries and not within a country (the pound in your pocket *is* worth the same); thus a pound now buys less of another currency rather than less tomatoes. Once it is de-fused and de-mythologized in this way then persuasive reasons must be found for treating this price change differently from any other price change. Thirdly, devaluation can, but not will, improve the Balance of Payments by lowering export prices since a commodity costing one pound now costs less in the foreign country. The final effect on the Balance of Payments depends on what it also does to import prices. The general point remains that if a Payments surplus is a policy goal then devaluation becomes a policy method worthy of consideration – to put it technically and not emotionally.

All this makes it surprising that, in some countries, such a gigantic brouhaha surrounds any devaluation. What are the difficulties?

Paradoxically, they arise from the very effectiveness of devaluation as a technique for allaying economic troubles. Lowering export prices can lead to more exports and to more employment as well. The difficulty is that this is a game that two can play. A chain reaction of rival devaluations is possible and was quite common in the 1930s. This is the first reason that many countries – particularly those heavily involved in international trade – avoid devaluation: they eschew the technique themselves because they do not wish it to be used against them in their turn. It is a kind of economic disarmament.

Stable exchange rates are valued in the second place because, in some countries, they are a matter of prestige. This is nothing to do with positive economics as such; the role of the economist is to point out the consequences of the normative decisions that make stable exchange rates a goal rather than flexible rates a technique.

Thirdly, stable exchange rates help to increase the level of international trade by bringing greater certainty to export prices than would exist in a world of fluctuating rates.

Fourthly, in the case of the UK at least, a stable exchange rate helps to bring international banking business which is itself a source of currency earnings.

Stable exchange rates offer then: prestige, a less uncertain world for international trade, and greater banking business as possible positive gains. They also help to minimize the chance of trade policies whereby one country transfers its payments, or employment, problems to another. The clear implication of all this is that countries will have to make internal adjustments, perhaps by credit and tax policies, to deal with this sort of problem. Choosing stable exchange rates as a goal of economic policy thus eliminates some of the policy options otherwise open. The present International Monetary Fund arrangements recognize both the general trend of this argument and the unreality of pretending that the price of anything is stable in three ways. Firstly, the IMF allows day-to-day fluctuations within 1 per cent of the official parity. Secondly, IMF arranges credit for countries with payments difficulties which, like the UK sterling difficulties after Suez, are considered both short-run and solvable, and which it would prefer not to be solved by exchange-rate changes. Thirdly, IMF allows devaluations in the face of a real long-run problem due to the unreality of a country's exchange rate, but the size of the devaluation is previously negotiated in order to minimize a string of retaliatory devaluations.

This at least is the principle; the facts of the matter do not always correspond too closely, for many small countries have devalued frequently and some major countries, like France, have rocked the boat by devaluing a number of times for unilateral gain, knowing that few countries would follow for fear of wrecking the system.

It should be noted that when credit is advanced to countries like the UK in payments difficulties, there is really no question of 'charity' for the other countries are simply 'paying' in one form to keep the system going – *it is in their interest*. Britain 'pays' to keep the system going in quite different ways – restrictive domestic policies and slower growth rates than otherwise and prestige losses – for those whose minds work along these lines.

A further point is that the UK and the US find it especially difficult to devalue because they are the major 'reserve currencies'. This means both that other countries hold some of their international reserves in dollars and sterling and that international trade is conducted in these currencies by countries other than the US and UK. This brings the psychic advantage to some of prestige and a possible economic advantage of greater international banking business. However, it also makes it difficult to devalue for a normative reason, and for a technical reason. If the UK and the US devalue, then the purchasing power of other countries' reserves is eroded and, in some sense, they have been 'let down'. Many people oppose UK devaluation on these grounds. The more technical reason is that if a small trading country devalues no other country is much harmed, but if the UK or US devalue then every country experiences the effect, and retaliation is more likely. Thus reserve currency countries can only devalue with the general agreement of others as in the case of the 1967 negotiations on the UK devaluation. Thus the paradoxical position is reached that some countries can break the rules of the game but the game still continues because, even in such a state, it is preferred to the game of freely fluctuating rates. Meanwhile, the UK and the US have to play strictly by the rules and are greeted with moral outrage if they have payments difficulties or seek to devalue.

It is easy to see why exchange-rate stability is a goal of government policy; it is difficult to see why it is held to quite so passionately, and even more difficult to see what alternatives are available until the entire international trade and monetary structure is re-cast.

We turn lastly to the Balance of Payments as a goal of government policy.

Whatever else the Balance is, in no sense is it a measure of a country's wealth as it is often interpreted to be; a moment's thought shows that some of the richest countries have the greatest trade difficulties, *eg*, the UK and the US. Nor should a businessman think of the Balance of Payments as some kind of profit and loss account. What then is the Balance of Payments? The answer must be preceded by some careful distinctions.

The Balance of Trade is the values of exported and imported goods. The difference is the 'trade gap', solemnly announced in television broadcasts and masochistically analysed in newspapers. *It is of no importance whatsoever in itself.* This is because it makes the pointless distinction of isolating goods from services. It is like a barber's shop valuing itself by its sale of razor blades and ignoring hairdressing charges. A measure of its importance is that in the past 169 years, the UK has had six Balance of Trade surpluses.

The Balance of Payments on Current Account is the values of exported and imported goods *and services*. If it is like anything, it more closely resembles an income and expenditure account. The current Balance never balances; if it were to do so it would be nothing more important than a curious coincidence.

The Balance of Payments on Capital Account is the current account brought into balance by capital movements – normally gold and currency reserve movements. The point is that there is no clear way of deciding what should be entered 'above the line' in the current account or 'below the line' in the capital movements. Once it is realized that at least some items can be put either above or below the line then it is clear that the existence and magnitude of any current surplus or deficit is attributable in part to the accounting conventions in use. In other words, the dramatically announced payment figures are not 'facts' at all. The size of the payments surplus or deficit is normative, for it depends upon what one chooses to put above the line and what below. This analysis does not tell us just how important the payments position is as a policy goal but it does warn against the most common misunderstanding.

Resource Allocation in the Economy

We must now pay some short attention to how scarce resources are allocated *between* firms, as well as *within* firms, so we may then see what economists have to say about this kind of resource allocation. Our attention thus passes from the private interest to the public interest.

Broadly, resources are allocated between firms by the price mechanism. The outline of the argument is that if the community wants a greater production of commodity A rather than commodity B this will, itself, lead to a higher price for A and a lower price for B which, in its turn, will lead to higher profits and so greater production from the A industry and the reverse in the B industry. The expansion of the A industry and the contraction of the B industry can be effected by the transfer of workers, capital, and land from one to the other in response to changes in the income levels available in the two industries. Such, in outline, is the market mechanism.

The first and obvious comment is that such an explanation is partial; many resources are allocated without reference to economic criteria at all. Churches allocate resources between worship and pastoral care on quite different considerations; not all firms take all the profit opportunities open to them, due to caution or incompetence; not all released resources in the contracting industry are employable in the expanding industry, and so on. However, something like the price mechanism does operate in some sections of the economy; what can be said about it?

This way of doing things is second nature in the Western world, and it is part of the layman's approach to economics that this system is 'good' and that government intervention is 'bad'. It should be emphasized that, though to some people the market mechanism is second nature, in fact it is rather a rare system both geographically and historically. It is not much used in sophisticated communist countries[3] nor in less developed societies, which means that it is unimportant in two-thirds of the world. It was not much known anywhere before the last few hundred years. What Western businessmen take for granted is really rather unusual in terms of human experience.

Many of the businessmen supporters of the price mechanism do not use it within their own firms where resources are allocated between departments by the board and where one departmental head cannot attract the secretary of another by offering her a higher salary.

Leaving aside political problems, what expertise can the economist bring to this way of running things? He is quite confident that such a system produces a sub-optimal resource use because it relies on money prices to measure costs and benefits. It is no surprise that the equi-marginal returns rule is not operated in a society where we have money prices that are a bogus measure of benefits and costs at the margin.

We have already seen that a firm is unlikely to know its

true costs, so it is unlikely that the price of its product accurately measures the true costs of production to be compared with production costs elsewhere. Even if a firm does know its true costs, its product price need not be related to that cost, as some firms are in a monopoly position and can charge prices unrelated to true costs. The economist can make an enormous song and clog dance about all this, but our purpose is not to show the layman the full battery of concepts at the disposal of the economist but simply to say enough to show that money prices are frequently a sham measure of costs.

A further mirror-image difficulty is that money prices also fail to measure true benefits. The value of a product to a consumer is shown by what he would have paid for it and this can be much greater than the price he did pay for it. The difference is called 'consumer's surplus' and where it exists then the market price underestimates true benefits.

Thus money prices can be poor measures of both real costs and real benefits.

In such circumstances it is no wonder that an economics course becomes a training in cynicism and that the economist is pessimistic about whether the price mechanism can allocate resources properly between firms or whether firms can optimally allocate whatever resources they get. It is important to say that similar difficulties and, in fact, others[4] arise in societies that do not use the price mechanism. That is, we are not saying both that the price mechanism is imperfect and that we know a better system, we are simply saying that the price system is imperfect.

One can rank it against other systems on political grounds but not on economic grounds; such systems can merely be compared.

The Private and Public Interest in Resource Allocation

There is still a widespread layman's viewpoint that what is good for a firm is good for society since the firm is part of society. 'What is good for General Motors is good for the USA.' It should be clear that this is erroneous. We have seen that firms, quite innocently, may allocate resources in a suboptimal fashion because their prices do not always measure true costs or benefits, or properly take account of externalities. However, there are more difficulties in the form of quite evident conflicts between private and public interests.

A few examples will suffice. A firm which engages in labour hoarding may be acting in its own private interests in that holding staff in a slack period, and so paying for a greater labour force than is then strictly necessary, may be cheaper than releasing staff and later recruiting again in a full order period. But it could be to the public interest that the staff work elsewhere in the slack period.

As we shall see in Chapter Three, savings which may be in the private interest of the saver may be at a level quite inappropriate for the public interest. Similarly, investment may be in the interests of a firm but against the interest of society if it contributes to inflationary pressures. Monopolistic practices are commonly thought to be in the monopolist's own interest but against the public interest. It is common to condemn trade union restrictive practices as a hindrance to economic growth but if they lead to better working conditions then the employees are better off; once again the interests of some do not coincide with those of others. Economic growth can take three forms: greater national output, greater leisure, or the same output with better working conditions. It is common not to realize that restrictive working agreements may make for improved working conditions and that this can be as valid a form of economic growth as the alternatives of more leisure or greater production. Restrictive practices are, nevertheless, sometimes seen as a barrier to economic growth rather than as a form of economic growth.

Once it is recognized that private and public interests do not always run in tandem then criticisms of government policies, because they hold back some activities of some firms, become invalid. It is not a criticism of public policy to show that it hazards private interests; this is not to say that all such policies are right but simply that they merit a more articulate attack.

It is necessary to take this matter of the public interest and resource allocation a little further, for use must be made of these concepts in later chapters. We will look first at the optimizing conditions and then at why they are so unlikely to be met.

These conditions – known as the Pareto conditions to economists – are further examples of the equi-marginal approach. It will serve our purposes to look at two of them. In terms of *what* is produced they require that the last unit of economic welfare, gained by an individual from a given product, shall equal the marginal welfare gained by all other individuals from all other products. This is unlikely both because it presupposes optimal income distribution and that the money prices give a true guide to alternative production forgone – otherwise a consumer cannot really know the alternatives available to him. If he does not know the choices in front of him he will only make the 'right' choice by accident. In terms of *how* it is to be produced the Pareto conditions require that each resource (land, labour, and capital), is allocated so that it increases the economic value of output equally at the margin in all uses.

These conditions are unmet in the real world because money prices do not show the value of true alternatives and, therefore, do not show true costs because the *real cost of anything is the alternative production forgone*. Money prices do not show true costs because of the monopoly element in some prices, the difficulty of marginal adjustments, and the problem of externalities.

Resources would be allocated in an optimal fashion if the prices showed us the value of alternative production at the margin, that is if *prices equal marginal cost*. The marginal cost is the difference made to total cost by the last unit produced. This is why, both in planning problems and in nationalized industries, the economist is much interested in setting prices equal to marginal cost. Although there are exceptions, a general rule of thumb can be that the public interest is maximized and resources are optimally allocated if all prices equal marginal costs. In later chapters we will look at the calculation difficulties but it should be admitted at the start that these are immense and sometimes give anomalous results – we can all see that sometimes an extra passenger will cost a railway company nothing more than the cost of printing his ticket, and that in other cases they would have to run an extra train.

Some implications of our argument can now be made explicit. If prices do not accurately measure costs or benefits then profits do not measure what production is 'worth' nor whether it is 'efficient' from the vantage point of the economist.

Profits are the difference between those benefits for which a firm can charge and those costs that it has to pay. These private costs and benefits do not have to be in the same ratio to one another as total costs and total benefits. Not only are some costs and benefits excluded from the profit figures but even those that are measured are inaccurately measured. Natural scientists are lucky in that 100°C. is a consistent measure at all times. Social scientists are unlucky in that £100 is an inconsistent measurement. We are thus brought to the conclusion that profits are important for all manner of purposes – they help to determine dividends, taxes, and future company policy – but they do not measure efficiency or economic value. *Unprofitable does not mean uneconomic and profitable does not mean valuable to the community.*

Under pressure, businessmen sometimes claim that though profits are not an accurate measure of efficiency they are a rough and ready one, implying that higher profits mean higher efficiency even if not in the same *proportions*. Even this cannot be allowed to pass; a loss-making operation is not always inefficient, it simply means that private costs exceed private benefits, but this carries no implications whatsoever about the relative magnitude of social costs and benefits. Similarly, to know that a process is profitable does not tell us at all whether it is efficient or worth while. This raises a presumption that there will be cases where it is important to measure these social costs and benefits.[5]

For these sorts of reasons, *it may be perfectly sensible to allow a nationalized industry to continue to run at a loss and for the Ship Industry Finance Board to give 'subsidies' to public companies – indeed it may be the only way to 'finance' economically worthwhile activities.*

Economic Growth and the Activity Level and the Firm

Having looked at some of the problems of resource allocation we can now turn more briefly to consider how economic growth and the level of economic activity modify the strategies of the firm – more briefly as these topics will be returned to in later chapters.

We need, first, to distinguish increases in economic activity from economic growth. Obviously, if unemployed resources become employed then there is an increase in output but this is also true of economic growth. The difference is simply that if it were always technically possible to produce the greater output without forgoing alternative production then it is an instance of increased economic activity, but if it were not technically possible to produce the extra output beforehand then it is an instance of economic growth. Much public discussion of economic growth, especially in newspapers and on television, is really about fuller employment. Both processes can, of course, occur at the same time and may be part of an increase in National Income.

How do changes in economic activity and the growth process affect firms? The brief answer is that since they change the firms' environment they alter the rules of the game. Once this is done, what was ill-advised may become well-advised and vice versa. For instance, economic growth can change the size and structure of the market so that a firm which could see its markets limited by the purchasing power needed to buy its products must revise its ideas. It was by this sort of reasoning that the British car industry embarked on expansive programmes in the late 1950s.

An increase in the market would be quite a different

matter; it would just mean more cars. Economic growth means that and differing proportions of different types of cars. However, changing the firm's scale and pattern of output means changes elsewhere. It may mean a different proportion of total spending on advertising as the new product-mix needs a different mode of selling; possibly long-established ratios between ordinary shares and debentures might be altered, especially if inflation is expected as this reduces the real burden of debt; as output increases so it may be possible to alter the proportions of labour and capital since automation is usually more available at greater outputs; as output increases so stockholding normally needs to increase less than proportionately – and so on. Economic growth alters the options open to a firm and in changing what it tries to do, it changes how it does it.

It is important to realize that though we can easily think of sequences of events which economic growth implies for a firm there are, nevertheless, no general rules. The effects will differ greatly from firm to firm and the consequences for a particular firm can only be predicted if details of its production costs and markets are closely specified. It is worth remembering that economic growth is not inevitably in everybody's interest – some industries, like bicycles, decline.

The way in which an increase in economic activity might modify a firm's policies is rather more explicit, though there will certainly be differences from case to case. If a firm goes from a two-shift to a three-shift system then it has altered the ratio between labour and capital and almost certainly lowered capital cost per unit of output. This probably means it should use more capital and this may alter its debenture policy, its training policy, and so on. The fact that capital and land is now getting a fuller use every twenty-four hours, because of a shift increase or because of greater capacity working from the same shifts, will probably mean that a new investment policy becomes appropriate, including, perhaps, a new depreciation policy and so on.

Enough has been said to show a firm cannot be myopic.

Its parochial internal problems may be pressing, genuinely important, and difficult to miss, but they are better dealt with if seen against those problems that are easy to miss until it is too late; the problems of the changing economic environment. No firm is an island; no firm can take the view that the closing of the Suez Canal affects oil companies but not itself. The economy is complex and interconnected, and what affects one part affects all others, some parts immediately and obviously and other parts much later and marginally.

Our last consideration is the nature of the changing environment; the aspects which interest economists are the changing resources and the changing objectives. The interdependence of the parts of the economy which we have emphasized should not be misunderstood; it is not like a car which depends upon a great number of components working simultaneously and will cease altogether at the failure of one part and which must be stopped completely to fit a new carburettor to give a better performance. The economy is more like a growing organism than a machine; in the greatest depression some parts have kept going, despite the failure of others, and an economy can grow to give better performances while it is still working.

The nature of change in the economic environment must be made clear. It is sometimes held that economics is the science of choice and that we must choose because some resources are fundamentally limited. This is fair enough in the short-run but is basically an unhelpful viewpoint. In the short-run there are limited resources but in the long-run this is not really the case – apart from eccentric examples invented for the purpose such as the difficulty of getting a million Yehudi Menuhins. But, for the most part, if we want something we can get it – whether it be computers, hospitals, food, or whatever. As the Americans have shown, if you want the moon badly enough you can buy it. There is, of course, a catch. We can have almost anything but we always have to pay a cost in the form of forgoing an alternative. To say we can have almost anything is different from saying we can have everything together at the same time. Put another way, the menu of economic possibilities is large – but we have to select from the menu.

The menu changes due to alternatives in the quality and quantity of resources and to changes in our objectives.

No firm should regard its resource availabilities as limited in other than the short-run. There is a stream of new capital goods and of savings available for equities and debentures – the firm's problem is to get its share. The supply of labour may be limited at a given moment but the less skilled can be made more skilled and labour can be geographically mobile – again only if the firm is willing to pay the price by allocating expenditure to training and so away from something else. The supply of land is fixed but in the long-run the planning authorities provide some supply of new residential land, and marginal farm land can have enough money spent on it to become better farm land. We can always have more of something, which is not to say we can have more of everything that is technically possible. And if we do have something extra then we must go without something else we already have or might otherwise have had for the first time. Similarly, what we wish to do with our changing resources alters with time. For instance, as we get richer we devote more resources to leisure, and less, proportionately, to the production of goods.

It is sometimes held that some resources are more truly scarce than others, that, somehow, capital is a more fundamental constraint on us than the other factors of production. This misses the point, which is simply that we can have more of almost anything but that always there is a cost in the form of forgoing choices.

A firm must learn then not only that it is not an island but that it must look in its own interest at the rest of an ever-changing economy. The firm must look outside itself more than occasionally otherwise it will deceive itself, it can take nothing for granted. Now you see it, now you don't.

References

1 This list is adapted from the Radcliffe Report, 'Report of a Committee on the Working of the Monetary System', Cmd. 827, 1959.
2 For a detailed treatment, see Mishan, E. J., *The Costs of Economic Growth*, Penguin, 1969.
3 See Chapter Seven.
4 See Chapter Seven.
5 See Chapter Four for an account of cost-benefit analysis.

Bibliography

The Firm:
Hague, D. C., Managerial Economics, Longmans, 1969.

The Economy:
Pen, J., *Modern Economics*, Penguin, 1965.
Schultze, C., *National Income Analysis*, Prentice-Hall, 1967.
Stewart, M., *Keynes and After*, Penguin, 1967.

CHAPTER TWO

The Resources of the Economy

In Chapter One we mentioned a concept central to economics, that of the allocation of scarce resources. Since a firm is involved in this process of resource allocation and since the nature of the economic environment is greatly determined by the quality and quantity of resources available, our next task is to evaluate the UK's resources. This chapter deals with the various forms of resource available to an economy, with the problems of measurement, and with one recent valuation; it continues by looking in more detail at labour, land, and capital.

By a resource (or Factor of Production) we mean any item, tangible or intangible, available to the nation for use in the process of production. Production is to be interpreted in its widest sense to include the provision of goods (*eg*, agricultural output, raw materials, capital goods, consumer goods, building, and construction) and services (*eg*, banking and financial services, the Civil Service, the professions, education, and health). A resource may be tangible, such as a deposit of coal, an acreage of agricultural land, a machine, a factory, or a bridge. On the other hand, it may be intangible, such as the training assimilated by a worker or the expertise of the London money market. What all resources have in common is that they can (but do not have to) contribute to the satisfaction of the wants of society and that they require initiative to be used and may command a price in use.* Trying to evaluate them makes it necessary to

* There are some 'free' goods, of which air is an example. Other factors may be virtually free to the individual firm, *eg*, technical knowledge published and not subject to patent restriction.

put them into categories so that we can discuss items which have similarities greater than their differences. We will use a conventional division into labour, land, and capital. The first two are obvious enough: capital requires some further explanation.

To an economist, capital means *real capital*, resources which have been produced and are available for further processes of production, *eg*, a machine, a factory, or a bridge. These are, of course, some of the examples of tangible resources just given; this emphasizes the point that we are concerned with real assets and not paper titles to them – which are the 'Issued Capital' of a public company. Thus, if we were valuing the factors of production controlled by a public company we would be interested in land, labour, and capital in our own sense of real assets. To include the Stock Exchange valuation of its shares would be to double count and hence inflate the valuation. Throughout this book capital will be used in this special sense. This follows the normal usage among economists, who distinguish between a consumption good as one wanted for itself, and a capital good as one wanted in order to produce further consumption goods.

Thus capital is anything which produces a stream of benefits over time. This would allow our three factors of production all to be considered capital. And so they are; as we said earlier we followed the categorization land, labour, and capital because it is conventional and convenient. The convention became widespread in the 19th century when land, labour, and capital (and their earnings – rent, wages, and interest) reflected distinct economic classes in England. These distinctions have become blurred. The stock of labour cannot be valued as 'pairs of arms' but in terms of knowledge and skills – a form of capital which is added to by education and training as well as by a health service. Land is important not so much for its 'original and indestructible powers' but for the investment of fertilizers, drainage, etc, which have gone into it and for the buildings constructed on it.

The distinguishing features of many of the natural resources of land are a growing public awareness that they are a national asset (*eg*, National Trust areas) and that the resources are consumed in use (*eg*, coal reserves). So, too, are other forms of capital but, unlike machines, natural resources are irreplaceable and the supply is naturally given. But this is not the same as being fixed for all time, since whether a natural resource is considered an asset at all depends on technical knowledge and prices. Thus, oil deposits were once considered a curiosity and most oil shale deposits are not valuable given the present world price of crude oil.

Although we have mentioned the valuation of resources this turns out to be largely an empty promise in the analysis which follows. Comprehensive official statistics for the nation just do not exist. There are, of course, physical measures for a sector (*eg*, steel capacity, acreage planted with barley) and financial measurements (*eg*, capital employed in the motor industry), but if we are asked for consistently valued stocks of assets for the nation as a whole we are rather in the position of the schoolboy asked to add two apples and three pears. He produced the correct answer of two apples and three pears.

The National Income Accounts, which we have, measure flows and not stocks.* As we shall see in the next chapter, this presentation reflects the approach of most current theories regarding the way in which the economy works. The absence of a balance sheet makes it impossible to shift our thinking about companies over to the economy. To think of the economy as 'Great Britain Ltd' is not only conceptually wrong, as we hope to show in the next chapter, but is impossible in practice because we do not have enough information.

* For those with no accounting knowledge at all, the relationship between flows and stocks in any economic unit can be explained as follows. Imagine a business worth (assets – liabilities) £100 at the end of a period. If, during the next period, it produced total revenues of £250 and incurred costs of £200 its profit would be shown in the flow accounts as £50 but could also be found by inspecting its balance sheet valuation which would be £100 + 50 = £150.

There are many difficulties in measuring the national wealth – we will take two examples. Firstly, it is often necessary to revalue the physical assets owned by firms, *eg*, undervalued freehold premises. Secondly, some assets are never valued at all, *eg*, what would be the value of the nation's police force? We know the annual money costs but not the value of its output, still less a balance sheet valuation.

It is likely that we will have official national wealth statistics quite soon. Up to now the only information of this type has come from research studies.

At present the only recent figures available are the work of Professor Revell[1] who has calculated a UK Balance Sheet at December 31st for the years 1957–61. What follows is based entirely on his study from which some figures for 1961 are reproduced in Table 1. Basically this consists, in its full form, of a list of stocks of Physical Assets, Financial Assets, and Liabilities analysed by 14 sectors of ownership (*eg*, personal, banks, non-financial companies, central government, external, etc, ignored here).

TABLE 1
UK Totals 1961

	£ billion
Physical Assets in the UK	84
Plus Financial Assets	145
Less Liabilities	116
EQUALS Total Equity	113
Less Share Capital	29
EQUALS Net Worth	84

Source: Revell, J., *The Wealth of the Nation*, Cambridge University Press, 1967, page 9.

Conceptually, 'Financial Assets', since they represent money claims and not real resources (*ie*, Factors of production) are cancelled out by liabilities, since the same value is used for an item when it appears as an asset and as a liability. 'Total Equity' consists of net assets after liabilities which could be appropriated by creditors have been deducted. 'Net Worth' reflects the difference between 'Total Equity' and the Stock Exchange valuation of shareholders' interest. This measures the extent to which some of the nation's wealth cannot be appropriated by any individual person. The cause of this quite large difference is that Stock Exchange market prices reflect a great many factors, *eg*, expected dividends, rather than simply the assets of the company.

This exercise does not, however, measure all of the nation's wealth since only 'valuable objects that economic units would regard as part of their wealth'* are included. Thus collectively-held wealth (*eg*, a market price valuation of the National Health Service as a going concern) is excluded, as are mineral wealth (*eg*, coal, oil, and gas reserves), and human capital (a valuation of skills and training embodied in the population).

Clearly the physical task of collecting the information for such an 'unofficial' statistic is immense, and its authors would agree that great inaccuracies are at present inevitable. Thus 'cash' is valued as a Financial Asset at £15,851 million but as a Liability at £18,209 million. Greatest use of such a balance sheet, as with National Accounting flows, will only come when up-to-date time series of figures are available and, perhaps, when international comparisons are possible. Nevertheless, some more or less meaningful statements can be made from our elementary presentation.

1 A figure of 'national wealth per head' can be calculated for the collector of 'pure' useless information, *eg*:

1961 UK £1,322
USA £3,409 (at exchange rate $2.80 per £)

2 The composition of the physical stock of assets in the UK can be compared with the USA, in the full balance sheet, part of which is shown in Table 2 below.

* Revell, page 19. There are remarkably few examples of omissions through practical difficulties of measurement, one such case is works of art and antiquities.

Some differences are explicable, *eg*, the very low proportion of land in the UK's physical assets. Others are at present unexplained, *eg*, the relative importance of 'Other Land and Buildings' and 'Plant and Equipment'. The higher proportion of 'Stocks and Work in Progress' in the UK, despite the existence of a more compact geographical market, relates to a possibly less efficient use of resources here. Techniques for stock control have advanced considerably

TABLE 2

Comparison of Components of Net Stock of Tangible Wealth as percentages, 1958

	UK	USA
Agricultural Land	3·0	6·1
Other Land	5·9	12·7
Dwellings	24·8	24·9
Other Buildings and Works	23·8	25·6
Plant and Equipment	24·2	12·1
Consumer Durables	4·5	10·8
Stocks and Work in Progress	13·7	7·9
	100·0	100·0

Source: Revell, *op. cit.*, page 295. (Totals rounded.)

in recent years in the USA and it has been suggested[2] that this represents a part of the American growth rate which could be followed here. Comparisons between output and capital employed are subject to many qualifications, but differences such as 'Stocks and Work in Progress' could help to account for the GNP/National Wealth ratio being 3·4 in Britain compared to 3·7 in the USA.

Table 1 does, then, provide some evidence, measured in terms of money, at market valuation, of our resources of land and capital. We will now turn to a more detailed examination of each factor of production in turn. For the most part we will be concerned with measurement in terms of physical quantities and an attempt will be made to describe national strengths and weaknesses in this way.

Labour

According to the 1961 Census, the population of Great Britain and Northern Ireland totalled 52·7 million, making the country the eleventh largest in the world measured in this way. Looked at as a resource we are interested not so much in the total, but in that part of the whole which either is, or potentially could be, employed. Table 3 gives figures which approximate to this total.

TABLE 3

UK Working Population – September 1968

		million	million
	Working population		
	male	16·3	
	female	9·0	
A.	Total		25·3
B.	HM Forces		0·4
C.	Wholly unemployed		0·5
D.	Total in Civil Employment (A–C)		24·8
E.	Employers and self-employed		1·7
F.	Total employees in employment (A–[B+C+E])		22·7

Source: *Statistics on Incomes, Prices, Employment and Productivity* No 29, HMSO, June 1969.

These figures are only approximate and may at the present time be misleading. The reason for this is that 'A' in Table 3 is in practice compiled from information concerning 'B', 'C', 'D', and 'E'. While 'B' and 'C' are known quite accurately for any particular week, 'D' is obtained from quarterly checks on National Insurance Cards. But 'E' can only be found from a Census of Production – of which the last was in 1966. Normally it is safe to rely on assumptions that this will not vary quickly. However, the introduction of SET and higher social security stamps have led to a great increase in workers becoming self-employed in such industries as building. Thus there has been a large and unmeasured (probably around ½ million) shift out of 'D' into 'E' which has affected the statistics at present in use only by reducing 'D'. As a result, we have currently no reliable figures for 'A' (the working population) and thus other statistics which use 'A' (*eg*, unemployment percentages, activity rates) are inaccurate.

These figures are an imperfect measurement of the size of the labour resource, not merely because of any statistical inaccuracy, but because an unknown part of the population not at present employed or available for employment (in the sense of being registered unemployed) could enter the working population if, for example, the school leaving age were lowered, retirement age postponed, or married women attracted into paid employment.* Whereas for a male over 15, and under 65, the choice is between being employed or unemployed (and in either case staying in the working population), for a married woman the choice may be to work or stay at home – since if paying the minimum National Insurance Stamp there are no financial benefits from registering at the local Labour Exchange and the woman may be uninterested in the job-finding aspects of the Labour Exchange service.

It may seem unlikely that the working population will be increased by changes of the first two types although, of course, as a proportion of the whole it may be changed by alterations in the age and sex structure of the population. It has proved very difficult to predict population changes; however, a general tendency for the working population/total population relationship to fall seems reasonable and a recent official prediction is that the working population will fall by about 3,000 *pa* during 1968–72 (compared with a rise of 187,000 *pa* 1960–66). This turn-around of nearly 200,000 workers *pa* results from lower birth rates in the 1950s and the current excess of emigration over immigration.

Changes in the working population which follow from alternations in the structure of the total population have two separate implications. Firstly, the labour market will tend to become more difficult for firms who rely on sections of the

* For most the alternative is to be a housewife. Obviously in such a capacity they contribute towards society's needs; an example of the arbitrary nature of our definition of 'working population'.

working population which will decline, *eg*, males under 20. Secondly, the 'burden' borne by the working population will increase; thus total population, 1967–72, is expected to increase by 2·6 per cent, but the 'over 65' and 'under 10' age groups, both of which make heavy demands on the social services, will increase by 8·4 per cent and 4·7 per cent.

One statistical attempt to measure the working population/ total population relationship is by the Employee Activity Rate.

$$\frac{A}{B} \times 100$$

where A = estimated number of employees,
B = employed population over 15 – including those not available for employment (students, housewives) and members of the Armed Forces.

EARs have been calculated for the Standard Regions of the UK. For males there are only small regional differences, whereas for women there appear to be large unused reserves of female labour, especially in the North, the South-west, and Wales. This seems to relate to the industrial structure of such regions, which has resulted in weaker traditions of women working than has been the case in, for example, the textile areas of Lancashire. Clearly there would be a problem of establishing the direction of the causal relationship here. Does the industrial structure which offers employment opportunities for women cause more married women to work or are there regional social characteristics – which, of course, may be associated with other factors such as male incomes – making female labour more willing to enter the working population in some areas rather than others? This could be important for a potential employer considering re-location; if he offers jobs for women, what sort of response will there be? The point has been disputed; but the most recent view is the one first stated, namely it is the structure of industry, and the jobs which are available, that determines the female Employee Activity Rate of a region.

TABLE 4

Regional Analysis of Unemployment percentages – April 1969

	per cent
London and South-east	1·6
Eastern and Southern	1·8
South-west	2·7
Midland	1·9
Yorkshire and Humberside	2·7
North-west	2·4
Northern	4·9
Wales	4·0
Scotland	3·4
Northern Ireland	7·1
National Average	2·4

– representing 558,000 unemployed persons.
Vacancies at this date were 303,000.
Source: As Table 3.

TABLE 5

International Comparisons of Unemployment percentages, average for year 1968

	per cent
Germany	1·5
Ireland	6·7
Italy	3·5
Japan	1·2
Netherlands	1·8
Norway	1·4
Sweden	2·3
UK	2·5
USA	3·6

Source: *Bulletin of Labour Statistics*, ILO, Geneva, 1969, Table 4.

The second point which affects the number of employees working is the extent of unemployment. Table 4 gives some figures for Great Britain while Table 5 makes some international comparisons. The extent of unemployment has become a goal of government policy (see Chapter One) and the overall forces which determine it are examined in the next chapter. Some features of the problem are, however, best dealt with at this stage.

We can distinguish various reasons for an individual being unemployed (in the statistical sense of being registered as available for work at a Labour Exchange) at any given point of time.

These are:

1 Personal factors, such as age, physical health, and attitude to work. This perhaps includes that popular group the 'workshy', but also includes those whose earning capabilities are below the subsistence social payments paid while unemployed.
2 Workers in seasonal trades who have difficulty in finding jobs out of season.
3 Those who are between jobs but will have little difficulty in finding other employment. Attitudes to the number of these 'frictionally unemployed' will depend upon the point of view. The individual unemployed may look upon the period as an investment of time spent in finding a better job – there is some evidence that statutory redundancy payments and earnings related unemployment pay have resulted in people spending longer on the unemployment register in order to improve their chances of finding a 'good' rather than just 'another' job. From the point of view of employers, increases in the frictionally unemployed represent increased labour turnover – and thus increased costs. From the point of view of economy as a whole, frictional unemployment may be seen as a response to, and a necessary part of, change in the environment.* Society must surely see some frictional employment as 'beneficial' in the technical sense of increasing total output, round pegs in holes of the correct circumference are likely to be better used resources contributing more to national product. This is, however, difficult to quantify. A simple net increase in labour productivity is not enough, since some industries have higher

* See Chapter Five for an analysis of government measures to improve labour mobility.

labour productivity than others (see Chapter Eight). The lesson to be drawn by the manager responsible for labour turnover is that national factors may affect the results of an individual firm, and that policies may be in his firm's interests but not society's.

4 Those who are unemployed as a result of change but whose skills are so specific or whose age is such as to make re-training difficult. Change in this context may be the replacement of a skill by technical developments or the decline of an industry due to changes in demand.

5 Those who are unemployed as a result of cyclical fluctuations in the overall level of activity in the economy (see Chapter Three).

Table 6 gives a categorization of unemployment in 1964. The analysis does not follow our 1–5 above; however, we have indicated a possible relationship against each column.

One important category included which has not been separately analysed by us is those unemployed for 'lack of local opportunities'. The causes of this regional unemployment are looked at later. Our analysis would include 'lack of local opportunities' as being the result of structural or cyclical forces. There is also some evidence of regional differences in the 'frictional' proportion of the total. Some regions seem to concentrate a given proportion of unemployment on to a smaller number of the unemployed.

To summarize our view of unemployment: we have a proportion of any unemployment total representing a resource not being used. To reduce this figure may involve the economy in costs, either direct, as in re-training, or indirect, as in the inflationary pressure generated by full employment. Some unemployment is unavoidable in our type of society, *eg*, physical disability, some, *eg*, frictional, is consistent with change.

Table 4, in addition to giving unemployment figures, also showed the number of vacancies in the country at that time. Such figures are often presented in association with the number of unemployed. If there is a near match between the two, it is sometimes assumed that the unemployment is not

TABLE 6

Prospects of Securing Employment – Registered Unemployment, UK 1964

	Code*	Men	Married Women	Single Women	Total
Should get work without difficulty	3	52,920	9,900	9,690	72,510
Will find difficulty in getting work					
Lack of Local Opportunities	2, 3, 4, 5	37,110	11,190	5,660	53,960
Present Qualifications, experience or skill are not acceptable to employers	4, 5	3,280	1,030	800	5,110
Will find difficulty in getting work on personal grounds					
Age	1, 4, 5	54,790	3,280	5,110	63,180
Physical or Mental Condition	1	47,960	4,020	8,060	60,040
Prison Record	1	3,760	20	20	3,800
Attitude to Work	1	24,380	2,750	2,170	29,300
Colour	1	2,190	640	470	3,300
Lack of English	1	900	60	80	1,040
Restriction on Availability	—	1,200	6,740	1,400	9,340
Lack of Financial Incentive	1	3,580	30	50	3,660
Non-members of Trade Union	1	360	—	—	360
Other Reasons	—	4,090	1,790	1,450	7,330
TOTAL		236,520	41,450	34,960	312,930

* Refers to categorization in text.
Source: *Ministry of Labour Gazette*, April 1966.

a problem – since a possible definition of full employment is where those seeking work equal the number of vacancies. We do not think that such comparisons are meaningful. Partly this attitude is the result of statistical shortcomings. Whereas the vast majority of those in the working population, but unemployed, will register as such – even though they may be unlikely to find work – there is no legal compulsion for employers to notify a vacancy. Indeed, where the labour market is 'tight' they may feel that to do so is a wasted effort. Unregistered vacancies are estimated to be often at least twice those registered. More important, crude national comparisons may be quite misleading.

Thus in April, 1969, national unemployment totalled 558,000 as against 303,000 vacancies. But the ratio varied widely with the occupation, sex, and age of the worker. Thus: 'Male labourers' 240,000 unemployed, 13,200 vacancies, and 'Female, food manufacturing workers' 287 unemployed, 14,411 vacancies. There were also considerable regional differences. Thus for 'Engineering and Allied Trade Workers':

	GB	London and South-east	Wales
Unemployed (thousand)	31·9	4·8	1·7
Vacancies (thousand)	26·2	5·7	0·7

Such differences, demonstrated by even the simplest analysis, show the limited value of unemployment/vacancy ratios as they can be obtained at present. This is not to deny their possible value. Explanations of why given vacancy figures may be associated with different unemployment rates will only be possible when more accurate vacancy figures and more up-to-date categorizations of both estimates are available.

So far we have concerned ourselves with the quantity of labour resources in Great Britain measured as numbers in the working population. This is an incomplete measure of the supply of labour effort for which we also require measures of length of time worked per person and intensity of effort.

Table 7 gives some international comparisons of hours worked. The figures are not offered as being particularly comprehensive or meaningful. They do show, however, that our working week is not greatly out of line with the supply of effort of some other major industrial countries.

We should, perhaps, mention another reason for a decrease in the supply of effort – annual holidays, our public holiday entitlement being one of the smallest in the world. Whereas, in 1962, 3 per cent of manual workers were entitled to more than two weeks annual holiday, by 1968 the proportion had risen to 44 per cent. The attitude of others to

TABLE 7

International Comparisons of average Hours Worked per Week in Manufacturing Industry for year 1968

France	45·3
Japan	44·6
Spain	44·1
Switzerland	44·6
UK – male	45·8
– female	38.2
USA	40·7
West Germany	43·0

Source: As Table 5.

the holidays at present received by the writers make us diffident on the subject – but we would point out that the wants of society may include leisure as well as other goods and services, and that this 'minor' change in weeks worked per year far outweighs the time lost by strike action. Quite rightly this is a comparison not often made, but it is worth making if the criterion remains simply hours of productive effort per year.

It may be thought that intensity of effort would be measured by labour productivity. Later in this chapter and in Chapter Eight we will explain why such figures may be meaningless and misleading.

Strikes obviously affect the supply of effort. What part do they play in the British environment? Public opinion stresses both the number of hours of production lost through strike action and that the small, unofficial strike, to which it is claimed our labour relations system is particularly prone, has particularly wide secondary effects.

Table 8 reproduces some figures from a recent study which show that, measured by 'number of working days lost per 1,000 employees', our record is very similar to most of the other major manufacturing countries – although very much 'worse' than West Germany and much 'better' than the USA. This study points out that all international comparisons are made difficult by differing definitions of a strike and the fact that some industries (*eg*, mining) are more strike-prone than others, but these represent different proportions of a country's industrial structure. On the whole, the figures

TABLE 8

Number of Working Days Lost per Thousand Employees. Annual averages 1964–6

UK	190
Australia	400
Canada	970
Denmark	160
Finland	80
Germany	Fewer than 10
Japan	240
Norway	Fewer than 10
USA	870
Belgium*	200
France*	200
Ireland*	1,620
Italy*	1,170
Netherlands*	20
New Zealand*	150
Switzerland*	40

* Figures unreliable due to obscure national definitions.
Source: *Is Britain Really Strike-Prone?*, Turner, H. A., University of Cambridge, Department of Applied Economics, Occasional Paper 20, 1969.

of days lost by strikes are more reliable than the number of stoppages. It concludes that the evidence for the allegation that we have a high proportion of small strikes is slight, that we are not unique in having a majority of unofficial stoppages and that there is no evidence that legal regulation, coupled with decentralized bargaining, would reduce the strike burden (*cf* GB with USA and Canada).

One difficulty in discussing the importance of strikes is that it is very difficult to calculate their effect in quantitative terms. Obviously the crude estimates often given in the popular Press are wide of the mark when 'value of lost production' is used as the measure. Firstly, if the figure is arrived at by dividing annual sales by the number of days for which the plant was stopped there is an implied assumption that the plant works continually at full capacity (virtually unknown) and that the lost output cannot be made up later. A similar assumption has been made when expected weekly output is used as the measure of loss. Secondly, the cost to the firm is not lost sales but lost profits – invariably a fraction of sales revenue. Thirdly, if lost exports are calculated on a basis of the proportion of annual sales exported, this assumes that the proportion of subsequent production devoted to the export market cannot be varied.

For these reasons, the usual measurements of the costs of strike action are quite misleading. We do not know of any successful techniques which have actually measured the economic costs of strikes to society in an economically meaningful way. But we suggest that the approach should be along the following lines where manufacturing industry is involved (there are obviously separate problems in, for example, dock strikes or where perishable goods are spoilt). Any buffer effects of stocks are ignored.

In the case where all firms involved are working at full capacity levels, then the lost production *is* equal to the output lost by the final manufacturer in a chain of production. But this assumes that the 'other industries' (*ie*, suppliers of raw materials and component parts) are not only working at full capacity but unable to divert their production elsewhere. In the unlikely case where all such production could be diverted, then the loss of output would be the 'Value Added'* by the firm.

Where the full capacity assumption does *not* apply, and lost output could be made up later, then the loss is the difference in present value† between the old pattern (*eg*, 4 in time period 1, 4 in time period 2, 4 in time period 3) and the new (*eg*, 0, 6, 6). (Whether this calculation is applied to sales, profits, or values added depends on the reasoning above.)

* See Chapter Three. † See Chapter One.

Measured in this way, the costs of a strike are likely to be very much less than is popularly believed. But there are other intangible costs which may, in the long run, be more important than anything discussed so far. This might be the case if potential, or actual, strike action hinders the application of new technology in an industry.

We are not claiming that our present system of collective bargaining, or the organization of the parties to a bargain, is perfect. We are saying that proponents of radical changes are theorizing with little evidence to support their cases. Finally, to put the whole issue in perspective, strikes in GB in 1969 (which has been the 'worst' year since 1957) account for less than one hour per head per annum of the working population – or about 2 per cent of time lost by sickness and injury. The facts are thus very different from the view commonly publicized in the Press.

The reply that strikes are not inevitable, and should be prevented, while sickness and injury are unavoidable, is incorrect. Few claim that the present level of strikes could not be reduced or, on the other hand, that all strikes can be prevented. Doctors and safety officers are clear that sickness and injury can be reduced. Like those who handle our system of labour relations, they require resources. Benefits involve costs; it remains unproven that resources could not be more effective in producing one benefit rather than another.

We now turn from considering the amount of labour resources of the country to some comments about its quality. We can pass over allegations about the 'natural' superiority (or inferiority) of the race in terms of intelligence, powers of leadership, etc, on the grounds that we know of no evidence on this matter. Health standards also affect the quality of the labour force as well as its quantity, *eg*, by improved life expectancy. International comparisons are difficult to make, but clearly we are healthier than the majority of the world's populations.

Education and training are clearly the two most important factors which affect the quality of the labour force. It is impossible to analyse the British educational system within the confines of this type of book. In addition to the lack of research into the economic costs and benefits of the system as a whole, the fundamental feature of any education system is its success in meeting the social objectives of the nation. Since the system is operated by the State to serve essentially non-economic objectives, we feel that it is essentially outside the manager's concern. On the other hand training *is* a matter of great importance to him – which we have considered in the second part of Chapter Five.

Land

In one sense, the quantity of this particular resource which is available is known and is fixed – the area of the United Kingdom is 60,297,000 acres or 94,214 square miles. If we expand the resource to include the natural qualities of the area, *eg*, climate and mineral deposits, they are by now largely known (but NB natural gas) or uncontrollable (*eg*, climate), or unexploitable with present technology (*eg*, shale deposits).

A potted geological survey of the United Kingdom seems out of place here; Table 9 analyses land use of the surface area. As soon as we start to analyse the natural resources of the economy we find that, like other factors of production, land and other natural resources can vary over time; for example, Table 9 would have been quite different 10 years ago (*eg*, barley 1957, 2,622,000 acres; 1967, 6,027,000 acres). The value of a particular form of land may change and, within limits, the proportion of the whole devoted to a particular use will vary.

Change in land use may be the result of new tastes – thus mountainous areas such as Snowdonia now have an amenity value whereas 200 years ago they were considered worthless marginal farm land and by contemporary standards not even beautiful. New products resulting from changes in tastes may enhance the value of particular sites – such as the recent revival in the use of china clay from Cornwall. New technology causes areas to be relatively more desirable than they once were. For example, nuclear power generating requires extensive sites with good load-bearing subsoil, close to the sea but away from a large centre of population.

Since agriculture receives considerable subsidies from the public sector (between £200–300 million *pa* in recent years) and since the form of the assistance and the level of subsidy per crop has varied over time, government intervention has changed land use by changing the relative profitability of various types of farming.

TABLE 9

Land Use in the United Kingdom, June 1967

	thousand acres
Wheat	2,305
Barley	6,027
Oats	1,012
Mixed Corn	88
Rye	11
Potatoes	708
Sugar beet	457
Fruit	227
Vegetables	409
All other crops	881
Fallow	229
All crops and fallow	12,354
Grassland	18,299
Total for crops and grassland	30,653
Rough grazings	17,639

Source: *Annual Abstract of Statistics*, 1968.

As demand for land with particular qualities fluctuates so we find changes in land use. Land is like the other resource with which we have dealt – labour – in that it is mobile between industries and between occupations. Unlike labour, it is not, of course, geographically mobile. This mobility is a reaction to, and a necessary part of, change in the environment. Government attempts to improve labour mobility are looked at in Chapter Five. We will now concern ourselves with the distinctive characteristics of land as a natural resource.

Public interest in the quality of the natural environment has recently increased. Since the 1947 Town and Country Planning Act there has been legislation which has sought to control land use on behalf of the nation. The demand for this particular interference with the free market mechanism stems from recognition of the amenity value of 'the countryside'. To the economist, this is one example of the distinction between private and social costs and benefits, the most obvious example of which are externalities of the type discussed in Chapter One. One example is the spoiling of the countryside in the past by the results of open-cast mining. It was in the community's interest that the land be restored but not in the commercial interests of the firms concerned.

But the problem at present is shortage of information regarding the costs – which are not merely regulatory legislation and higher local prices, but often very indirect. Thus some proportion of the present subsidies to farmers may be justified by their contribution to keeping the countryside in the form that town dwellers appreciate. But we do not know how much society is prepared to pay for amenities.*

Land as a national resource also varies in value not simply because it is mountainous and infertile but as a result of its position within the country. So while Britain is fortunate in having few inaccessible or mountainous areas (except perhaps the Scottish Highlands), it still has areas which for other reasons are more or less 'advanced' than others.

'Advanced' in this context does not refer simply to the degree of industrial or agricultural development but to differences measured in terms of unemployment, levels of income, and growth of living standards. Regional differences in unemployment percentages were given in Table 2. Other figures for incomes are shown in Table 10. The broad outlines of 'the regional problem' are common knowledge, in the sense that most are aware that unemployment is lower and

* Discussions about land use are frequently in terms of the 'loss of good farming land' resulting from a new housing estate, airfield etc. It was shown some years ago, however, that the market value of the produce grown in an average domestic garden need only rise to £20 *pa* for the urban yield per acre to equal that of best farm land.

incomes higher in, say, London than Northern Ireland. Regional problems have many implications for the manager; they not only affect location decisions but may, for example, shape marketing strategies and determine personnel policy. In this chapter we look at the symptoms of the problem and consider government techniques of intervention. In Chapter Seven we turn to 'regionalism' as a form of economic planning.

TABLE 10

Regional Variations in per capita *Incomes, before Tax*

Standard regions	Index of average *per capita* incomes, 1959–60	Index of average *per capita* incomes, 1964–5	Index of growth of *per capita* income, 1959–60 – 1964–5
South-east England	115	115	121
East Anglia	82	86	124
West Midlands	108	108	101
East Midlands	97	95	95
York and Humberside	98	98	101
South-western	86	89	111
North-western	97	96	95
Northern	86	81	77
Wales	83	83	101
Scotland	87	87	100
Great Britain	100	100	100

Source: 106th and 109th Reports of the Commissioners of HM Inland Revenue.
Reproduced from Stillwell, F. J. B., 'Location of Industry and Business Efficiency', *Business Ratios*, Winter 1968.

Before outlining some of the causes of the problem and indicating possible courses of action we must first define a problem area. Three types exist in this country:

1 Agricultural areas suffering from depopulation through migration, high unemployment, perhaps underemployment, (*eg*, agricultural proprietors who are not fully employed all year) and low living standards. Examples are the Scottish Highlands, parts of Ireland and mid-Wales.

2 The 'Depressed Areas' of the 1930s which rely heavily on once staple, now declining, industry (*eg*, coal, ship-building, cotton textiles). Problems here include heavy structural and cyclical unemployment and migration in the better qualified part of the population. Examples are South Wales and North-east England.

3 Congested areas with high population densities and low unemployment percentages, *eg*, London and parts of the Midlands.

In the case of the first two categories, it is not necessary for there to be an absolute worsening of conditions, either in terms of employment or living standards, for there to be a regional problem. It is sufficient merely for the gap between the problem areas' level of attainment or rate of change to have dropped further behind the national average – or the situation in the most favoured area – since this tends to be the criterion used by the population of such regions.

The mechanism whereby these regional differences arise is imperfectly understood. Much of the theory in this field concerns itself not so much with the region but with explanations for the pattern of industrial location. Three points are emphasized in this type of theorizing, although none are held to be necessary or sufficient to explain location. Firstly, that industries may be heavily influenced by transport costs, particularly the cost relationship between moving inputs to the factory and moving outputs to the market. Secondly, that industry may be attracted to existing centres of population since proximity to a market and easy access to a labour force may be considered most important. Lastly, that industry may be attracted to areas where similar industry has already located itself – where it can take advantage of auxiliary services, specialized labour, etc, which have already been developed.

There are, in addition, some intellectually fascinating, highly sophisticated theories (which in the last analysis have little practical application) concerning the development of regions as a whole. Given various assumptions, one will even demonstrate how specialized centres of manufacturing activity will develop on an area where, initially, natural resources were evenly distributed. Another shows how, once we have an urban concentration, there will be a tendency for it to grow in relation to the surrounding area as it acts as a magnet for the more advanced technology and the more ambitious businessmen of the region.

No single theory available to the social scientist explains the reason for regional differences though all those mentioned provide certain valuable insights. One important feature, however, is that there will be a strong tendency in industrialized economies such as ours for patterns of land use to be perpetuated. We are now at the stage in the discussion when the normative questions can be asked, *eg*, should we interfere with 'natural' economic development?

Four arguments can be put forward to say that we should – and these have been used to justify the panoply of controls and incentives which now exist in this country.

Firstly, one may argue that, in the same way that the redistribution of income from rich to poor is an economic goal of society (see Chapter One), so we should attempt to eliminate differences due to an individual's birthplace. A possible alternative, to encourage migration, is dealt with in Chapter Five. Perhaps most people who live in the better-off regions would be in favour of increasing living standards in poorer areas – assuming that they could accept that migration does not represent a complete answer. For evidence of this, consider Eire where the population has halved over the last 100 years but where emigration still exceeds the natural population increase, unemployment is higher and earnings lower than in Great Britain.

The problem becomes more difficult, however, when it is realized that redistribution in this way involves costs. These costs may be measurable, such as financial payments in the form of higher investment grants or the Regional Employment Premium. Other costs are not measurable and it may even be disputed whether they exist. An example would be reductions in national growth caused by shifting resources to less favoured regions. Even when financial costs can be measured, no convenient measure exists for comparing the increase in welfare obtained by those being helped, with the loss in welfare of the helpers.

The second argument for intervention rests on the case that, left to itself, the market mechanism only provides a suboptimal pattern of industrial location and that this can be intervened in, with advantage, by the government. To provide examples of location decisions which are not in the best interests of society is not enough; it also has to be shown that it is likely that changing the pattern will be an improvement. If we turn the argument round and seek to claim that businessmen should be left to make location decisions without interference, two weaknesses in the case appear. Firstly, surveys show that many firms do not seek the best, but merely a satisfactory site, for a new factory. Often they do not even seek the necessary information to give themselves a chance of selecting the best site. Although consultants are sometimes used to research and report, a far more typical approach is to seek to expand next door to the present premises or to search for the nearest suitable site. There is ample evidence for the existence of a considerable number of 'footloose' industries which have no strong preference for a particular area – many of whom have responded to government incentives.

What we have really got so far is that in many cases little harm may be done to the private sector by forcing on it locations which it would not otherwise have used or by providing incentives which artificially cheapen 'problem areas' compared with other sites.

Our second criticism of the free market system is more fundamental in that it provides a positive argument for intervention. It turns out to be yet another example of the differences between private and social costs Let us consider the case of a firm considering a move into the London conurbation (the third type of problem region). Quite properly, it will move if this increases its profitability – which assuming no change in revenues will come from lower costs per unit of output sold – perhaps, in our example, lower transport costs through closeness to its markets and lower labour costs through increased labour productivity. Such a decision, however, inevitably affects society. Any increase in the region's labour force caused by the move will increase congestion; the factory may be built on what was once land with amenity value. It is very difficult to put a price on the reduction of welfare felt by all those affected in this way – but this does not mean that it is unimportant. Society may also have to meet tangible increased costs as a result of the influx into the region. Examples are the increased pressure on the educational and social services of the district and the increased burden on the commuter transport system – whether in the form of more crowded trains or more congested roads.

It must be repeated that there are two quite separate effects operating here. The first is an increase in congestion in the sense of too many people and buildings in a given area and is very subjective. The second is an increase in costs forced upon the community.* There are, in addition, the intangible costs of policy objectives which may have to be forgone, *eg*, those associated with inflationary pressure from fully employed regions, mentioned below.

Thus there is no necessary reason to suppose that the best size of conurbation from the point of view of industry is the best for society as a whole. There is evidence to this effect; the figures vary with the study, but the results typically show that with over 200,000 persons the average costs of providing social services (in the widest sense) rise, although urban areas much larger than this provide ever increasing opportunities for external economies of scale to firms.

Our next reason for wishing to interfere with the market mechanism may be simply a belief that although it reacts to

* The authorities *could* decide to hamper growth by not providing extra facilities and thus letting the service deteriorate or they may be incapable of meeting the financial costs. Allegations of this sort are for some reason printed continually in London evening papers. The unfairness of such tactics is, of course, that it may not be the newcomers who are penalized.

change, the reaction time is too slow. To go back to an earlier example, 100 years of falling population in Eire is too long a time span in most people's opinion. This must remain a theory, there is no way of demonstrating the case one way or the other, partly because over long time periods the conditions which caused the initial reaction may themselves have changed and partly because regions change dynamically. Thus the knowledge that the population is likely to fall will affect industrial development in Eire and thus perhaps initiate further emigration.

The third argument for a regional policy is that, since 'problem regions' of the first two types have higher unemployment percentages than others, a specifically regional policy may be successful in using resources which would otherwise be unemployed. The calculation that then remains to be done is whether the costs of increasing the occupied population in this way are less than the resulting benefits.

The last basis for regional intervention is that in this way the overall growth rate of the country can be improved. There are many conflicting theories of growth but there is a considerable body of professional opinion which holds that one of the reasons for lack of growth in this country has been unacceptable inflationary pressure associated with conditions of full employment (see Chapter Three). Nation-wide policies may therefore be constrained by full employment problems, which when analysed geographically can be seen to apply only to certain areas. Thus a regional policy may allow, say, Wales to continue growth which would otherwise be held back by anti-inflationary policy in London, the South-east, and the Midlands.

We now turn our attention from the reasons for a policy to the policy itself. We have spent, in terms of the overall constraint of the length and balance of the book, a considerable time outlining the basis for intervention in a belief that most managers have been affected at some time by some aspect of the problem and the policy. The policy – in the sense of current government action – can easily be found out. It is the reason for any action at all which may seem dubious.

TABLE 11

Regional Policy: Principal Incentives received by Firms

Type of Project	Incentive							Remarks
	A	B Building Grant*	C	D	E	F	G	
	Invest-ment Grant	25 per cent	35 per cent	Opera-tional Grants	Training and SETP	Regional Employ-ment and SET Premium	Board of Trade Loan	
Factory for incoming manufacturer	X	P	X	P	X	X	X	C and D almost certain if SDA chosen
Expansion by existing manufacturer	X	X	P	P	X	X	X	C and D possible if firms at present outside SDA, and expansion is in different products and uses different management

Factory for newly formed company	X	X	P	P	X	X	X	Quality of backing crucial
New or expanded service centre and warehouses		P	P		P		X	B or C depends on employment factor
Incoming office employers		P	P		P		X	No problem if employer building own offices
New hotel		X					X	Special Assistance available
Incoming capital-intensive manufacturing project	X	P		P	X	X	P	Scaled down B and D likely a ceiling on G

NB X = probable eligibility P = possible eligibility
SDA = Special Development Area. Set up as a result of the White Paper on Fuel Policy 1967. Originally areas likely to be especially affected by mining closures, they have been extended so that Millom (Cumberland) is now included.

* Preference shown for developments where male jobs exceed 50 per cent of total.

Source: Adapted from *A Guide to Government Incentives for firms expanding in North-east England*, North-east Development Council, April 1969.

Current policy takes three forms:

1 Encouraging labour mobility (see Chapter Five).
2 Interfering with the re-location decisions of firms.
3 Using public money to finance regional development.

Table 11 details the benefits now received by firms in Development Areas. As established under the Industrial Development Act 1966, they now include 40 per cent of the surface area of Britain and 20 per cent of the population. They comprise: all Scotland except Edinburgh, all Wales except small areas on the south and north coasts, the Northern Region, Merseyside, and Cornwall and North Devon in the South-west.

As can be seen, they are if anything wider than the 'problem areas', types 1 and 2, mentioned earlier.

In addition to these inducements, the Board of Trade controls expansion through new factory building in London, the South-east, and the Midlands, for instance by requiring that an Industrial Development Certificate must be obtained for all but the smallest building. IDCs are commonly withheld in such areas in order to force relocation into a Development Area.

The attempt to lessen congestion in the London area by withholding permission to build has been reinforced by the Location of Offices Bureau. Set up in 1963, the LOB encourages movement away from Central London, largely to other sites in the South-east, by publicizing and acting as an information centre for the advantages of such moves. By March 1969 it could claim that it had helped firms in this way to the extent of over 50,000 jobs. On the other hand, the annual movement has been declining from a peak of 14,000 in 1967–8, largely as a result of the impact of stringent controls over new office building in the whole South-east region. This is claimed as evidence of muddled thinking by those who think only of the LOB's objective of movements out of Central London; on the other hand it may be sensible to seek a more even spread of office employment over the South-east as a whole, while still restricting growth of the region *vis-à-vis* the rest of the country. Regional policy is full of such apparent (and also real) contradictions.

Table 12 shows the cost of assistance.

TABLE 12

Government Expenditure on Financial Assistance to Manufacturing Industry in Development Areas

	1968–9 (estimated £ million pa)
A. Non-recoverable items	
Grants under the Local Employment Acts	24
Investment Grants (20 per cent differential)*	75
Regional Employment and SET Premium	124
Training Grants	3
	226
B. Recoverable items	
Loans under the Local Employment Acts	15
Building of Advance Factories for sale or rent	13
	28
Total estimated expenditure	254†

* Firms actually receive £150 million, but half of this would be available were they located outside a Development Area.

† Comparable totals in earlier years are: 1961, £56 million; 1963 £22 million; 1965, £66 million; 1967, £108 million.

This analysis ignores spending on the social infrastructure (*eg*, housing, amenities, and communications) designed to make Development Areas in general more attractive locations.

Source: Kan, A., and Rhodes, J., quoted in letter to *The Financial Times*, April 16th, 1969.

To sum up; we now have varied tools of regional policy which affect the availability and price of natural resources in this country. New building may be subject to control, either to preserve particular zones of land use under the Town and Country Planning Acts, or to force re-location in Development Areas by IDCs. The value of certain parts of the land surface may be increased deliberately by devoting resources to them which the market would not provide. The justification for the policy is a mixture of social and 'economic' factors which do not all unequivocally call for the same type of action. The effects of the policy are difficult to estimate, there is some evidence that unemployment figures have tended to draw closer to the national average, particularly since 1965, and some data show that some problem regions have grown at a faster pace than the national average. Measured in terms of lowered unemployment as an objective, the problem regions now seem to be the North, Wales, and Northern Ireland – the last named with unemployment rates still 2½ times the national average.

But, apart from short-term noticeable social effects, *eg*, the provision of a job for an unemployed man, we are still not sure, in theory or practice, of the results of any action taken, or indeed of the cause of the problem. All countries have their regional problems, ours are small measured in terms of a dispersion from a national average. That the differences exist should not be surprising, for similar differences exist between nation states. Perhaps the analogy should be drawn more often between aid to underdeveloped countries and aid to problem areas.

All this is highly relevant to firms for whom re-location is a possibility. Regional economics has become very popular academically in recent years; the 'spin-off' from this interest has been a growth of regional information which, properly analysed, can be of great value to industry.[3]

Capital

We now turn to the third resource available to a country – capital, the definition of which was given in the introduction to this chapter. We can describe the forms of capital but, for reasons already explained, we are unable to give any reliable estimates of the present size of the resource. All is not lost, however, since a feature common to most forms of capital is that their value is reduced over time.* This may be because they are worn out in the process of production (*eg*, the wear

* The exceptions are largely those cases where shifts in demand cause the market price of the asset to appreciate. Nevertheless there is usually the contrary tendency at work. A house may rise in price, but it is still deteriorating.

on a machine tool), deteriorate simply through the passage of time irrespective of the amount of use which they receive (exterior dilapidation of a building), or lose their commercial usefulness by obsolescence in the face of technical change (windmills being replaced by coal-fired steam power, coal-fired electricity generating stations being replaced by natural gas).

The amount of reduction in value of an asset (a piece of capital) per year is here termed depreciation. Theoretically, this describes the same process which is allowed for in accounting by a depreciation calculation. To the layman a half-understood view of (accounting) depreciation is that it describes the technique of putting on one side in a business, assets which will be available for replacing the unit of capital when it wears out. This is a mistaken view of the accounting term, which merely consists of reducing profits by the annual amount of the reduction in value of the assets used. The provision of resources within the firm to allow a replacement to be bought is a quite separate business. The process is recognized by accountants and economists; differences do arise as the result of accountants often using conventional formulae (or simply seeking to obtain the maximum allowance from the Commissioners of Inland Revenue), whereas economists, who rarely have actually to do it, remain true to the concept, *ie*, that depreciation is simply the difference in value of any asset between the first and last dates of the period being considered.

The three causes of depreciation mentioned above provide the motive for capital owners to maintain their stock of capital by replacement. In addition, of course, enterprises which are growing may require increased stocks of capital. We do have measures of new capital formation, whether to replace depreciated assets or increase the stock of capital, and this compensates for the lack of a total valuation.

Any addition to the stock of capital is termed investment by the economist and figures are available for the proportion of the annual production of the country allocated in this form. In addition, it is sometimes possible to analyse the

Gross Investment figure (*ie*, the total) into Net Investment (gross less depreciation) and the difference between Gross and Net, which represents an increase in the stock of capital.

What all investment has in common is a diversion of resources into producing capital goods rather than consumption goods and services. Thus, if the total flow of goods and services from the economy cannot be increased, increased investment must be associated with decreased consumption in that year, unless imports have risen to a greater extent than exports.

TABLE 13

An Analysis of UK Gross Investment
Annual percentages GNP average, 1950–62

	per cent	
Construction	4·1	
Machinery and Equipment	7·6	
Non-residential fixed Equipment	——	11·7
Residential Construction		3·0
Inventories		1·0
Net foreign lending		0·5
Total Gross Investment		16·2

Source: *Why Growth Rates Differ*, Denison, E. F., The Brookings Institution, 1967, page 118.

Table 13 analyses Gross Investment on an annual average for the years 1950–62 into its major parts. This emphasizes that while the major part of investment is 'non-residential fixed equipment' (machinery and factories), house building is also an important part of the whole. This distinction becomes important later when we are considering ways in which total investment can be increased. 'Inventories' refers to increases in stocks. The measurement is an imperfect one; but clearly the holding of stocks of finished goods and work in progress is as much a capital resource as plant and equipment. In some cases it may be possible to substitute one form of capital for another. Thus steel stockists may see as alternative forms of investment to cater for increased sales, either larger stocks or mechanical handling equipment enabling them to turn an existing stock level over more quickly. 'Net foreign lending' refers to investment overseas.

There has been considerable Press discussion recently about the quantity of investment by the United Kingdom. In particular, unfavourable international comparisons have been made and it has been alleged that this provides a major explanation for the slower rate of UK growth compared with other European countries and Japan.

Table 14 shows some international comparisons and also shows growth rates in investment for the period 1950–62. On this basis the British performance is poor – but even an elementary analysis such as this should take certain other factors into account. Firstly, the figures are percentages of GNP, and since GNPs and GNP per head (see Chapter Three) both vary widely between countries, the percentage quoted represents very different absolute values of investment and investment per person. Secondly, since the price of capital equipment varies internationally (so that for non-residential fixed capital equipment UK and European prices are about one-third higher than in the USA) two percentages giving the same money total would represent different physical quantities.

Thirdly, measurements of gross investment mix together depreciation replacements and additions to the capital stock. The former are likely to be higher the greater the stock of capital at the start of the period being examined. Thus if Country A had a stock of 100 units and 400 GNP and Country B had a stock of 1,000 units and 400 GNP, using a depreciation rate of 10 per cent would result in 10 units (*ie*, 2½ per cent GNP) in 'A', and 100 units (*ie*, 25 per cent GNP) in 'B'. The figures also need to be interpreted with the nature of the capital stock in mind, thus two countries may have the same capital stock but if in one a high percentage is in factories and houses, which typically depreciate slowly, depreciation may be smaller.

Fourthly, one may distinguish the source of the investment, *ie*, public or private sector (see also Chapter Four).

TABLE 14

Total Gross Investment as a percentage of GNP
An international comparison, ranking by annual average 1950–62

	per cent
Norway	28·4
Netherlands	26·0
Germany	25·9
Italy	21·8
France	19·5
Belgium	19·2
Denmark	18·8
US	18·4
UK	16·1

Source: as Table 12, page 118.

Growth Rates of Gross Non-Residential Fixed Equipment.
Ranked by annual average increase 1950–62

	per cent
Germany	5·5
Denmark	4·8
Norway	4·2
Netherlands	4·2
France	3·6
Italy	3·6
US	3·7
UK	3·0
Belgium	2·9

Source: as Table 13, page 138.

This type of detailed analysis on an international scale takes place only intermittently; Table 14 enables international comparisons to be made for the period 1950–62. The figures are gross but it is not believed that deductions for depreciation would greatly affect the ranking. This same study[4] also contains 1964 estimates for the *stock* of 'enterprise* structures and equipment' per civilian employed.

Denison concludes that 'the average British worker is supported by an exceptionally small amount of capital'.[5]

Why should this be the case? What effect may this have upon growth? Any answer to these questions requires a brief discussion of the process of investment.

* Includes government enterprises but not social capital.

Private sector investment, it can be assumed, will only be undertaken with the objective of increasing profits. This may be in the positive sense of increasing the stock of capital to either provide for growth of sales or produce the existing output at lower cost. Alternatively, investment may be necessary to keep pace with competition so as to maintain profitability. The expectation is profit, although the resources once used remain committed, whether or not the decision was commercially sound.

TABLE 15

Enterprise Structures and Equipment per Civilian Employed 1964. USA = 100

Canada	111
Norway	92
Belgium	67
Denmark	66
Netherlands	62
France	54
Germany	52
UK	44
Italy	36

Source: as Table 13, page 170.

Government-directed investment is not necessarily subject to the same test, firstly because it will take account of social as well as private costs and revenues, secondly because it may be used as a technique for meeting other objectives of the economy as a whole (see Chapter Three). Housing investment is also rather a special case; if only because it is the only major form of investment carried out by persons as opposed to firms.

Private Sector investment therefore depends upon such factors as:

1 The technical efficiency of the capital good under consideration. Thus the purchase of a machine in the UK may not be worthwhile now whereas a more efficient machine available in another country may stimulate investment there and the development of a new machine may stimulate investment here.

2 The availability and cost of borrowed finance for the purchase of the machine. More will be said about this in Chapter Five. The evidence is that shortage of funds has not generally been a constraint in the UK.

3 Future trading conditions influencing the likely sales success of the output of the investment. It is a reasonable hypothesis that the fluctuating, 'stop–go' economic progress which has been more characteristic of the UK than other countries may have reduced the incentive to invest here.

4 The effects of managerial efficiency. This may take two forms. Firstly, that British companies may not be seeking greater profits with the same diligence as those of other nations and may thus not have the same stimulus to invest. There is no hard evidence, in general, either way for this view although comparisons between US subsidiaries and their UK competitors in this country seem to show that US firms are more effective. There is evidence that the *rate of return* on investment in Britain is lower than in other countries with whose *rates of investment* we are often unfavourably compared.

Secondly, the inefficiency could be in the techniques used to evaluate an investment opportunity – so that firms 'don't recognize a good thing when they see it'. There are two pieces of evidence which support this point as having an effect on investment and thus the stock of capital in the economy. In the past, considerable tax incentives have been given for investment (see below page 68) the effectiveness of which has been reduced by many firms taking investment decisions on a pre-tax basis and thus producing misleadingly pessimistic results. New techniques of investment appraisal such as Discounted Cash–Flow, now gradually spreading through industry, on the whole show investment to be more worthwhile than the older, cruder methods.

Let us suppose that we can accept these indications as evidence that we have invested a smaller proportion of GNP than we 'ought' to have done What will be the result? Looked at very simply, an increase in the capital stock reduces the volume of consumer goods and services available now, but (it is hoped) will increase the flow at a later time period. How much later is important, as is, for what sacrifice. Perhaps most of us accept the 'Protestant ethic' and believe that we should sacrifice our present pleasure for our future good. But hedonism is a possible alternative attitude and not all of Aesop's fables have universal validity in an industrially developed society. To choose to increase the growth rate is a value judgement.

Our continued low rate of investment may have reduced our growth rate. On the other hand, the relationship between investment and growth is complex. Two equally plausible theories are possible. The first would be that an increased stock of capital goods causes growth by virtue of the more valuable increased flow of goods and services in the future. The second would claim that a growing economy provides a very strong incentive for firms to invest in order to take advantage of the large markets which can be expected. Both are equally plausible and it has not yet been possible to quantify their relative importance.

One relationship with growth (in the sense of higher incomes per head of the population) which has been demonstrated, however, is that over a long time span, say 50 years, for an advanced economy, technical change is a far more important source of growth than capital stock per head. This does not mean that rates of investment are unimportant: if it were true that little could be done about varying the rate of technical change then investment would remain the important variable. There is also a relationship between investment and technical change in the sense that the faster the turnover of the capital stock the more quickly new techniques can be embodied in capital resources (and not only new techniques, more investment would give the opportunity for capital resources to be moved geographically into better locations). What should be clear is that the investment/growth relationship is not as simple as it is often assumed to be. It does open the way for such theories as that which states that population growth (faster in most European countries than here) is important since this is a variable outside our circular relationship which would encourage investment.

However insecurely based on experience, the promotion of investment has now become a popular technique for solving our economic problems. Probably it would be *the* panacea but for the pandemonium that surrounds the Balance of Payments results. The government has a variety of techniques available for promoting (or retarding) investment. Two of them, monetary and fiscal policy, will be discussed in Chapter Four.

During the period 1961–5 a complicated system of initial allowances and investment allowances operated, with special regional differentials, which enable firms undertaking investment to have their tax liability reduced. The effects of the system were difficult to calculate but seemed, in terms of allowances against gross profits, to be more favourable than, for example, in the USA.

Since 1966 a system of *investment grants* has been in force which, where it operates, replaces the above measures. (Thus investment which does not rank for a grant may still attract initial allowances of 15 per cent for buildings, 30 per cent for plant.) Investment grants are a cash repayment of 20 per cent of the cost of investment undertaken by manufacturing and extractive industries only. Thus the system discriminates between industries; it also discriminates between types of investment (*eg*, vehicles, office equipment, and loose tools are excluded, there are special provisions for ships and computers) and depends upon location (the rate is increased to 40 per cent in Development Areas).

Although comparison is difficult between the old and new systems, most economists were optimistic about the grants scheme for the following reasons:

1 The size of the benefit, *eg*, £20 per £100 invested, was obvious to the potential investor. It did not depend on profitability or the firm's tax liability.

2 Quick repayment of the grant meant that the benefit

was certain, *ie*, it could not be varied by subsequent legislation.

3 Varying the repayment period, *eg*, between 6 and 18 months, provides a delicate way of varying the strength of the policy.

4 The assistance was selective both in terms of the industry and the nature of the investment. This selectivity could easily be changed as and when required.

5 Regional differentials were incorporated from the outset.

6 The total size of the incentives, in terms of grants paid, meant assistance as good as in any other major country.

Most commentators have been disappointed by the results of the new policy (*eg*, investment never met official 'National Plan' targets). To the surprise of many economists there has been considerable opposition by industry to the grants scheme – tax allowances seem, from the CBI's view, at least, a preferable alternative. Possibly this stems from an illogical distinction between having a tax bill reduced as opposed to a cash grant – the first being viewed as a reduction of a burden, the second as a subsidy – firms, as well as individuals, it would seem, dislike a 'dole'. Perhaps also the basic technique has suffered by association with the administrative problems caused by the discriminatory features of the scheme. There have been legal arguments about what constitutes the manufacturing sector and classificatory problems about certain types of plant. Lastly, some managers object to all qualifying investment being subsidized whether or not it turns out to be commercially profitable.

Why are we, as a nation, such poor* investors? Possibly as a result of managerial inefficiency in some cases (see Chapter Eight). Very likely as a result of a not very remarkable recent growth coupled with stop–go fluctuations. Perhaps simply because we prefer present consumption to future greater consumption (see Chapter Four).

* A value judgement again!

Conclusion

We have tried to make the most important relevant points for the manager about the resources of our environment. It is impossible to produce practically useful measurements for any individual manager – since clearly what he wants is detailed information which cannot be given in a general book of this type. In any case our objective is to broaden a manager's horizon by, firstly, making him aware of the fundamental forces which determine the resources of the whole economy and, secondly, warning him against the limitations of the measurements which he may meet.

We have seen that to produce a 'total' for each is statistically impossible and not very rewarding because in every case their quantity and quality can be varied. One word of caution – where this type of classification is involved it is very easy to become blinkered by the classification itself. Thus, when in the previous section we wrote about Capital and Investment, it should be remembered that our analysis did not always deal with all types of investment and that some expenditures, such as the running costs of the Health Service, have the qualities of an investment goad, even though in the figures for analysis it is considered an expense and not a piece of capital.

Finally, what is important in our environment is not merely the amount of each type of resource, but the relative quantities of each. We frequently compare some of these ratios internationally, *eg* population per square mile, capital per worker. By treating population as the fixed component we talk about 'over population' and 'under capitalization'. These may well be useful ratios, but a major problem arises when we calculate efficiency, treating one factor as fixed (and forgetting differences in national ratios) to produce 'output per man' 'bushels of wheat per acre'. More of this in Chapter Eight.

The firm is, then, in an environment which greatly affects its behaviour but the difficulty is that the stock of resources cannot be reliably identified or evaluated; the relationship between the firm and its environment is thus exciting and ambiguous – a point which will be re-emphasized when we turn to the problems of the employment and output levels.

References

1 Revell, J., *op. cit.*
2 Denison, E. F., *op. cit.*
3 Stillwell, F. J. B., *op. cit.*
4 Denison, E. F., *op. cit.*
5 Denison, E. F., in R. E. Caves, *op. cit.*, page 272.

Bibliography

This chapter is not well served by available reading matter. Parts of the contents are dealt with in:

Brown, A. J., 'Regional Problems and Regional Policy', *National Institute Economic Review*, No. 46, November 1968.

Caves, R. E., and Associates, *Britain's Economic Prospects*, George Allen & Unwin, 1968. Denison, E. F., 'Growth and Efficiency', Part 3.

Denison, E. F., assisted by Poullier, J-P., *Why Growth Rates Differ: Post-war Experience in Nine Western Countries*, Brookings Institution, 1967.

McCrone, G., *Regional Policy in Britain*, Unwin University Books, 1969.

Revell, J., *The Wealth of the Nation*, Cambridge University Press, 1967. For a brief account of this work see: Revell, J., 'The Wealth of the Nation', *Moorgate & Wall Street Review*, Spring 1966.

Stillwell, F. J. B., 'Location of Industry and Business Efficiency', *Business Ratios*, Winter 1968.

CHAPTER THREE

How the Economy Works

Having gone some way to describe the economy, our attention now turns to its level of activity; for the firm, however parochial the viewpoint of its managers, affects this level of activity and this, in its turn, helps to describe the world of the firm and, thus, the limits within which it operates. The firm, however unwillingly, however marginally, influences the economy, and the economy influences the firm, and so on, in a two-way interplay like an echo-effect. But it is not like an ordinary echo between cliff faces that slowly dies away; the reciprocal economic dependence is a continuous business, for the resources and objectives of both the firm and the economy are always changing and starting off new sequences.

It is worth emphasizing that just as the firm may neither know, nor care, how it affects other parts of the economy, so decisions taken elsewhere, which affect the firm, may be unrelated to the firm in any specific way. The firm is concerned in the sense that it is affected, but not in the sense that the decision maker considered each firm involved. A government's anti-inflationary policy can damage a firm's market although the firm had little or no part in the inflation; the same could be true of Balance of Payments or exchange-rate policies; a government's employment policy can place the firm in a tighter labour market in which wage costs rise and it must consider changing its labour/capital ratio (degree of automation), its geographical location, and so on. In other words, a firm simply seeking to maximize its profits can produce far-reaching effects, including some far from its thoughts; policy decisions elsewhere which are perfectly well designed for their immediate purposes can greatly modify the firm's environment and neither party need be aware of the other.

We have, so far, looked at the way in which the government and the firm, unknowingly, affect one another. It is also true, of course, that one firm can unwittingly alter the cost patterns or market possibilities of another; the discovery of North Sea Gas or the technical progress that results in new production methods can provide examples of this sort of thing.

In economics we seldom know the full consequences of our actions, and so it is particularly important to know that we do affect one another without knowledge or intention in a kind of economic blindman's-buff and, also, to know as much as we can.

What do we mean by the level of economic activity? An intuitive rule-of-thumb guide to the amount of activity is the employment level. In broad terms, full employment indicates a high level of economic activity and vice versa. As might be expected, it is not as simple as this; first of all an employment increase of a given percentage can mean a production increase of a quite different proportion, due perhaps to technological conditions, and, even if the production increase is in the same proportion, there is no need for living standards to rise in the same way, as the population which has to share this output may have altered in size.

Not only do employment changes not measure economic activity at all accurately, but it is even possible for output (and living standards) to alter without employment changing *at all* as, for instance, if a given working force works overtime or more efficiently. However, it remains generally true and useful, particularly in the short-run, to think of employment and economic activity as being intimately connected. For one thing it is realistic, since for social reasons the one aspect of economic activity that consistently attracts attention is (full) employment.

A second aspect of the activity level that attracts attention is the physical output of real wealth in the form of goods and services. In some ways, to economists, this is a more useful guide to the level of economic activity and it, like full employment, manifests itself as a goal of government policy – economic growth. This physical output of real wealth is, of course, a mixed bag measured in different units – hundredweights of minerals, bushels of grain, gallons of milk, thousands of medical consultations and so on – it is measured by statisticians in the common units of money prices and known to economists as Gross National Product. If GNP is shown as higher in one year than in another this means a real increase in production, and not the same production at higher prices; that is, GNP is normally measured in constant prices as from a year that should always be shown with the statistics. There are similar concepts to GNP such as Gross Domestic Product, National Income, at market prices and factor prices, and Net National Product, but the differences are not germane to our present interests.

It is helpful to note, at this stage, that the National Income can be conceived and measured as the money incomes, the money expenditures, or the money value of production. This is because every income has a matching expenditure and that expenditure must be the value of the production. This straightforward proposition is an essential step in explaining the level of the National Income as well as in its measurement.

All three of these approaches are used in estimating the National Income; a particular problem in the output valuation is to avoid double counting as, for instance, in adding together the value of the steel in a car and the value of the car. If these products are made by different companies we are interested only in the 'value added' by the car company and not in the price of its product.

GNP is not, of course, the National Income, since it is a gross figure which ignores the depreciation of capital to produce the output.

Domestic expenditure at market prices is greater than GNP at factor cost in that it includes expenditure on imports which are not part of domestic income and the tax element in prices. These must be excluded to obtain GNP at factor cost, while subsidies and other countries' spending on our exports must be included to reach a figure showing the incomes to the factors of production.

The figures given in Table 16 are estimates. They are the results of National Income Accounting conventions which, for instance, value most products on their prices, but some, like the National Health Service, on their costs; a further convention is to ignore unpaid household services which are undoubtedly products; as a result, a shirt washed by a housewife is excluded, but one washed in a laundry is included.

It should be noted at this point that an increase in GNP does not have to mean economic growth; it may simply mean a greater production due to a fall in unemployed resources.

There are a few points to be made before we consider the determination of the National Income.

Firstly, an increase in National Income does not necessarily mean an increase in economic welfare or 'happiness'. This is partly because GNP is a measure of material goods and services; one does not have to be a mystic to feel that there is much else to life – the visual amenities of a town as well as the number of houses, the quality as well as the quantity of food, the pleasures of family life, poetry, music, and so on. All these things can generate 'psychic' income but they are outside the scope of a restricted notion like GNP. The inequality between changes in GNP and economic welfare is partly due to externalities; as more transistors and more supersonic aircraft mean more people having to listen to them and, partly, to the more explicit costs of economic growth like the unemployment in the contracting industries and regions, only some of whose resources can be transferred to the prosperous sectors, and like the consumption forgone in the present by some to invest and produce higher living standards for others in the future. The level in the British GNP shown in Table 16 are thus a real gain but they must

TABLE 16

Scheme of National Income Accounts 1968

To illustrate the Relationships of the Different National Income Concepts

	By Expenditure (£ million)		
1	Consumers Expenditure	27,065	
2	Public Authorities Expenditure	7,702	add
3	Gross Domestic Capital Formation	7,798	add
4	Value of Physical Increase in Stock and Work in Progress	204	add
5	(1+2+3+4) *Total Domestic Expenditure at Market Prices*	42,769	result
6	Exports and Property Income from Abroad	10,670	add
7	*Less* Imports and Property Income from Abroad	10,679	subtract
8	*Less* Taxes on Expenditure	6,960	subtract
9	Subsidies	886	add
10	(5+6+9−7−8) *Gross National Product at Factor Cost*	36,686	result
	By Incomes (£ million)		
11	Income from Employment	25,267	
12	Income from Self-employment	2,840	add
13	Gross Trading Profits of Companies	5,117	add
14	Gross Trading Surplus of Public Corporations	1,352	add
15	Gross Trading Profits of Other Public Enterprises	111	add
16	Rent	2,359	add
17	(11+12+13+14+15+16) *Total Domestic Income before Depreciation and Stock Appreciation*	37,046	result
18	*Less* Stock Appreciation	650	subtract
19	Residual Error	129	subtract
20	(17−18+/−19) *Gross Domestic Product at Factor Cost*	36,267	result
21	Net Property Income from Abroad	419	add
22	*Gross National Product*	36,686	result
23	Capital Consumption	3,375	subtract
24	(22−23) *National Income*	33,311	result

TABLE 16 (cont)

Scheme of National Income Accounts 1968 (£ million)

Gross National Product by Industry (including Depreciation and Stock Appreciation).

Agriculture, Forestry, and Fishing	1,127
Mining and Quarrying	687
Manufacturing	12,527
Construction	2,456
Gas, Electricity, and Water	1,288
Transport	2,266
Communication	799
Distributive Trades	4,082
Insurance, Banking, and Finance	1,206
Ownership of Dwellings	1,801
Public Administration and Defence	2,258
Public Health and Educational Services	1,818
Other Services	4,731
Less Stock Appreciation	650
Residual Error	129
Gross Domestic Product at Factor Cost	36,267
Net Property Income from Abroad	419
Gross National Product at Factor Cost	36,686

Source: *National Income and Expenditure*, HMSO, 1969, pages 3 and 13.

not be mistaken for an unalloyed increase in economic welfare.

Secondly, it must be said that the international comparisons in National Income shown in Table 17 are tentative exercises. There are too many differences in statistical practices between countries for these to be anything other than a (good) rule of thumb.

A third point is that increases in money income must not be mistaken for increases in real income. A natural scientist would not measure a hundred yards with three feet to the yard in the first fifty and then two feet to the yard in the last fifty. This is what happens when GNP is measured in different years with different price levels. GNPs should be constructed with an index number, itself an imperfect measure, or with prices as constant from a given year. One of the great distinctions between the economist and the layman is the latter's faith in the measuring power of money.

A fourth warning remains before we can pass to the determination of the National Income. This concerns the fallacy of aggregation.

TABLE 17

Growth Rates of Real National Income (Total, per person employed, and per capita per annum)

1955–64

	National Income	National Income per person employed	National Income *per capita*
Belgium	3·5	3·0	2·9
Denmark	4·8	3·5	4·1
France	5·0	4·7	3·7
Germany	5·6	4·3	4·3
Italy	5·4	5·4	5·7
Netherlands	4·3	3·1	2·9
Norway	3·9	3·7	3·0
United Kingdom	2·8	2·3	2·1
United States	3·1	2·0	1·4

Source: Denison, E. F., *Why Growth Rates Differ*, Allen & Unwin, 1968, page 18.

It is a common enough idea that a society is not simply the aggregate of its members – that a crowd at a Nazi meeting, or a football match, can take on characteristics different from the persons in the crowd, or that somebody in a concert hall will applaud a performance that would be heard undemonstratively on a record in his own home. Chapter One of most textbooks in statistics contains a warning on relying on a sample of one, against thinking that because a penny comes down heads the first time that this affects the second toss of the coin. We are all properly wary of arguing from the particular to the general, of judging a nation by a person or a barrel by an apple. These dangers take a particular form in the case of economics where it is hazardous to argue from the individual, or the firm, to the economy. What is a sensible policy for a firm with a particular problem need not be sensible for an economy with the same problem. This is because an economy is not simply the sum of its component parts; thus a firm seeking to increase the numbers it employs might be well advised to cut wages, but a general cut in wages would not raise the general employment level (see later). So no economic argument should ever start in the form that what is true for one is true for the nation. An argument which is sound for the individual's case must be re-thought before it is used in another context.

Apart from this general warning against aggregating the economic behaviour of a community from that of its members, we will be content with one example. Laymen are sometimes concerned about the size of the National Debt on the analogy that a larger personal debt means a lower future living standard while the debt is repaid. The National Debt is in no way similar. In Britain it is a collection of UK government debt such as Savings Certificates, Treasury Bills and government bonds but the saving grace is that these are largely owned by British nationals. In other words, the National Debt is the exchange of money for paper certificates by private individuals while the government receives money for the certificates; later on the government redeems the certificates and exchanges money for them. The point is that all this simply amounts to the redistribution of money, and conversely paper assets, within the nation and, of course, the nation is neither richer nor poorer by all this. Even if the debt is repaid from taxation the economy is neither richer nor poorer since the taxation falls on the same group of people. The National Debt can, of course, make individuals within the society richer or poorer according to how it is spent and redeemed but it does nothing to the wealth of the community as such. The only part of the National Debt which is analogous to the private debt is that part which is foreign-owned. The National Debt is an internal transaction which does not impoverish the nation any more than a family is poorer if a member borrows from the housekeeping.

An outline of the recent structure of the National Debt is given in Table 18. Laymen are often disturbed by its size – which happens to be of the same magnitude as a year's

National Income. This is partly because of the fallacy that it is like an ordinary debt and partly because of a less widespread, more sophisticated, but equally mistaken notion that the National Debt transfers a burden from one generation to its successors. At a given time, the nation's real wealth can be allocated to public rather than private use, by means of government debt-financed expenditure, but its size is in no way affected; at a later time a different volume of wealth can be allocated from public to private use by government interest or redemption payments. Once again the nation's wealth is no different in size. There is no sense in which (domestically held) debt is a burden. A nation is made neither richer nor poorer if the National Debt is one-tenth of the GNP or ten times the GNP.

Of course, in explaining that it is not a debt we have pointed out the way in which it is important; it is simply a device for channelling resources into the public sector, and the economist's role is to analyse the economic and social differences between debt-financed expenditure and tax-financed expenditure: an economic difference is the opportunity cost in civil servants needed to collect a given sum of tax revenue against those needed to administer the same sum of debt; a social difference is the greater scope for specifically relating taxes to abilities to pay.

It is this lack of overlap between the economies of the individual and the firm with that of the nation, that has divided economics into micro- and macro-economics. It is macro-economics which offers an explanation of the determinants of the overall level of activity. This kind of analysis is also known as Keynesian economics after its main founder, J. M. Keynes.[1]

Once we ask ourselves the question what are the determinants of income it soon becomes apparent that we must rephrase the question because income is always associated with expenditure. Nobody can enjoy income without somebody else spending; rent incomes can only be generated by paid rent bills, wages that are income for some are wage expenditure for others and so on.

TABLE 18

Estimated Distribution of National Debt, March 31st, 1968 (£ million)

	Total	Treasury Bills	Total	0–5 years to Maturity	Over 5 years and undated	Non-marketable
Official Holdings	*9,168*	*2,378*	*5,924*	*1,328*	*4,596*	*866*
Non-Official Holdings of Public Bodies	190	—	190	48	142	—
Total Banking Sector	3,403	662	2,679	1,987	692	62
Other Financial Institutions	5,263	6	5,245	806	4,439	12
Total Overseas Residents	5,891	2,299	2,103	926	1,177	1,489
Total Other Holders	9,732	110	5,703	1,946	3,757	3,919
Total Non-official Holdings	24,479	3,077	15,920	5,713	10,207	5,482
Total Debt	33,647	5,455	21,844	7,041	14,803	6,348

Source: *Bank of England Quarterly Bulletin*, 1969, Vol 9, No 1, page 60.

Our question thus becomes: what determines expenditure?

The first, but not the only, determinant is clearly income: we would expect most expenditure to, in its turn, be preceded by income. We are hovering on the statement that income depends upon expenditure and expenditure depends upon income. This looks ridiculous but, in fact, it is quite helpful and an economy where income and expenditure were identical *would* be in a stable equilibrium. Once we have, in this way, specified the conditions for an equilibrium we can soon see the rather more important conditions for disequilibrium, for fluctuation is characteristic of most economies. Thus the concept of equilibrium is an important theorizing technique, for by seeking equilibrium we can soon see implications which explain variations in employment, output, prices, and income.

The determinants of the level of economic activity are soon identified if we ask ourselves not what determines expenditure, but what factors determine different kinds of expenditure and, also, when will income and expenditure be unequal?

Clearly income exceeds expenditure if *savings* occur (it makes no difference to us what form the savings take – whether they are in a bank or in a mattress, but it very much matters that the savings are genuinely 'withdrawn' from the expenditure flow so that money, for instance, saved by one person and lent to another, who then spends it, is not savings in our sense).

This brings us to an important point. When we postulated a simple economy in which income equalled expenditure we really meant consumption spending – that is on goods and services wanted for their own sake: food, fuel, clothing, and so on. So if income exceeded consumption spending, and there was no other spending, then the saving withdrawal would decrease the expenditure flow, incomes would fall, and so would the level of economic activity, and employment. But there *is* another form of expenditure – *investment* spending on capital goods.

Investment spending is regarded as an 'injection' which is the exact counterpart of the savings 'withdrawal' from National Income so that it has the opposite effect on the National Income level.

Clearly if these two are equal then the National Income (Y) is in equilibrium, but if $S < I$ then Y rises and vice versa. An important point is that there is nothing in the world to make the two equal at the same moment in time as the savers are not identical to the investors and the factors which govern saving and investing are not closely related.

We now have the Trojan Horse into the main structure of the Keynesian analysis: we simply look for further pairs of injections and withdrawals and inquire whether they are autonomously equal.

Another clear gap between income and expenditure is *taxation* and the countervailing injection is *government expenditure*. A government can easily run a budget surplus or deficit (decreasing or increasing the National Debt) and so there is no need for these two to be equal.

A similar pair of withdrawals and injections is provided by *imports* and *exports*: imports are expenditure within a nation that generate income in another country and not within the economy responsible for the expenditure flow, and exactly the opposite is true of exports. The most cursory examination of newspapers indicates that imports and exports are not automatically equal. There is nothing to make any particular withdrawal (W) equal its counterpart injection (J) nor is there anything to make the total J equal the total W.

We can now appreciate two important but straightforward propositions: the level of economic activity depends on the level of expenditure; changes in economic activity depend on inequalities between income and expenditure. If $W = J$ then National Income is stable, but this is unlikely, if $W > J$ the National Income falls, and if $J > W$ it rises.

This line of analysis leads to the conclusion that the level of economic activity should be up and down like a fiddler's elbow. It is a matter of common observation that though the national economy fluctuates it is not quite volatile in that sense. This is for two reasons: firstly, the factors governing consumption and investment spending are usually fairly stable in the short-run and, secondly, there is a built-in device whereby the economy is always adjusting towards an equilibrium National Income.

It takes the following form. If, for instance, $J > W$, perhaps because $I > S$, then the National Income rises and, *at these higher income levels more is saved so that the savings are brought towards investment in magnitude.* A similar argument is possible if the National Income is either rising or falling due to any inequality between any, or all, injections and withdrawals. Obviously our next problem is why, if equilibrium is reached, it proves to be a short-run stability.

The answer is that the very process of National Income change generates new patterns of spending during the time necessary for the income change to be effected. Thus, at some moment $I_2 > S_1$,* this excess of injection over withdrawal leads to a rise in income, but instead of savings rising to the S_2 level and thus producing an equilibrium $S_2 = I_2$, it is possible that, by that time, investment is at a level I_3, or for that matter I_1, or, even, that savings rise to S_3 instead of S_2. This will become more apparent when we have examined the factors governing consumption, investment and government spending and savings, taxation, imports, and exports. A further reason is that investment tends to vary in a manner that destabilizes the economy due to two relationships called the accelerator and the multiplier that we will come to later.

It is important to note, at this moment, that the economy is unlikely to autonomously produce its goal of full employment since (any) stable equilibrium National Income does not have to be a full employment National Income. If this is among society's goals then the government needs a policy that will operate on variables like investment and consumption to manoeuvre them to the levels necessary to

* What follows, the investment level I_1 is taken to be less than I_2, which is less than I_3. Similarly for savings.

achieve the ends. We can get a National Income equilibrium without getting full employment; we do not, therefore, want any equilibrium, we want the full employment National Income equilibrium.

A firm which finds that the State's tax, interest rate, or credit policies modify its own position, and stop it from doing what it would otherwise do, often talks of the stifling of initiative or effort, as though this in itself demonstrated that the policy was wrong. This is based on the fallacious assumption that the interests of the individual, the firm, and society all overlap. Such firms are simply finding that higher or lower levels of spending are required of them *in the public interest.* It is more proper to recognize these conflicts of interest as extraordinary and difficult rather than to engage in the self-righteous outrage and parochial viewpoints which tend to greet government agencies like the Board of Trade when it refuses factory building permission, the National Research Development Corporation when it refuses to back particular inventions, or the Treasury when it limits the spending of private tourists abroad.

We must now turn our attention to the factors which govern the various injections and withdrawals so that we may properly appreciate fiscal and monetary policy in a later chapter, and that we may see how volatile these are and how little economists know about them. We are interested because these make up the expenditure which becomes the revenue of firms and because they are the means whereby the firms unwittingly modify their own environment.

A résumé of economists' attitudes on these variables would take the following lines:

1 Both consumption and savings are held to depend mainly upon income in the form that the higher the income the higher the consumption absolutely, but the lower proportionately once saving has started, and the higher the income, the higher the savings, both absolutely and proportionately. There is, however, some doubt about whether the evidence means we should think of 'income' as absolute

income, or whether it is relative income or expected income that is important for consumption/savings decisions. 'Relative' income means income relative to family, colleagues, and neighbours. 'Expected' income means anticipated income in the forseeable future. These important distinctions need not detain us here. We are moving towards the unhelpful proposition that in order to get a particular level of income we must alter the spending/saving ratio and to do this we must alter incomes. There is not very much hope here. However, consumption and savings do respond to other factors apart from income but these are generally non-economic factors such as spending from a given income varying according to age, family size, and occupation. This information, too, is not much help to the policy maker.

In general, consumption and savings do not much respond in the short-run to income changes due, say, to tax changes, though they show more response in the long-run. It is implicit in all this that the economist has little sympathy for the layman's faith in interest rates as a means of changing savings. Interest changes may not alter the level of savings but they may alter the form – perhaps from the GPO to Building Societies. Although it is reasonably true that consumption spending is stable in the short-run, particular forms of consumption spending can vary while the general level is steady. Apart from seasonal changes, these movements tend to be due to changes in fashion or production conditions so firms cannot be too sure that even short-run consumption is safe for them, while for the government, consumption is a stable factor – which is convenient if it wants stability but difficult if it wants changes.

2 Investment is held to depend on the interest rate on the one hand and the return on investment on the other. 'Return on investment' is a clear enough term intuitively, although there are rival measurements in the trade which need not delay us here. It would appear that investment is a more hopeful target for government policy. The interest rate is reasonably open to government control and so, to some extent, is the return on investment. There are, however,

three difficulties: firstly, rising interest rates may affect other things besides investment in capital goods, such as house mortgages and the price of gilt-edged securities, and the government may not welcome these side-effects. Secondly, although interest rates affect investment there is plenty of evidence that their effect is small, partly because there are few investment projects undertaken where a one or two per cent interest change can halt the project. Thirdly, investment responds to interest-rate changes slowly as most projects take some years and may be better continued than abandoned in such circumstances. Interest rates, then, are a somewhat blunt weapon, more applicable to new investment than continuing investment.

The return on investment depends on the price of the capital assets – lorries, machine tools and so on – and the expected revenue stream they will produce. Subsidy and tax policy towards capital equipment and profits-tax policy are the obvious techniques here. Directly operating on the price of the capital equipment has given some modest results but they have varied between sectors and from time to time, so that it is difficult to be sure of the magnitude of the results of a particular policy. But profits-tax policy has shown some quick results so that company investment has offered an interesting mechanism for modifying investment spending in the direction favoured by the government.

In general, investment fluctuations seem to be explained by technical progress and the state of the Trade Cycle. To the firm this means that investment is rather variable and unpredictable; to the government it offers little hope of controlling investment, for technical progress is not open to reliable prediction and to know that investment depends upon the Trade Cycle is to recognize that it is high during prosperous booms, when it offers few problems so far as output is concerned, and that it is low during depressions just when it is 'better' high.

It is a matter of common observation that output, employment, and prosperity fluctuate more in investment goods industries, like shipbuilding and heavy engineering, than in consumption industries like food canning or hairdressing. The multiplier and the accelerator help to clarify this difference and also why the process of National Income change tends to lead to further changes instead of being finished once and for all.

The *multiplier* predicts that a given increase in investment will lead to an increase in income greater than itself. At first sight this is anomalous, as expenditure equals income, but there is no difficulty once we introduce time into the process for we are interested in income flows not wealth stocks. If a company spends £1 million on investment, that is £1 million of income to engineers, but the point is that they spend most of it, say £900,000, making a total income of £1,900,000; the recipients of the £900,000 spend most of it in their turn and so on. Thus, in time, the total effect on income far exceeds the initial investment. Not only are there investment multipliers like this, but also trade and government multipliers, all of which means that expenditure changes have long-range and profound effects on income.

The *accelerator* helps to explain investment fluctuations by postulating that investment increases depend on the rate of change of income: the faster the rate the greater the investment change. But the rate of change of income varies as we have seen and so in consequence does investment. A further difficulty is that investment spending is made up of a demand for extra investment goods and of replacement demand. Now this replacement demand is not itself stable for it must partly echo previous new investment fluctuations themselves connected with past income fluctuations. (Accelerators are, of course, capable of far more precise and complex formulation.)

We see then that investment is an important part of the destabilizing process in the National Income level, that it is so important that the government cannot leave it alone but it finds its behaviour difficult to predict and most awkward to manoeuvre to the level that the government requires. Meanwhile, firms in the investment goods industries are particularly open to change.

3 Imports and exports depend partly upon production patterns – countries obviously import what they cannot produce themselves – and then on the domestic and overseas price levels, for the same or similar goods, and the currency exchange rates. The government tries to control the domestic price level and encounters both difficulties and some success as we shall see, and it can control the exchange rate for some reasonable period of time if it is determined to so do and willing to forgo alternative goals. However, the main motive of the government in intervening in international trade is its concern with the Balance of Payments and the exchange rate for their own sakes rather than their role in determining the National Income. From the firms' point of view, the effect is much the same: society requires them to behave in a particular way to imports and exports in the public interest. It uses tariffs, subsidies, and quotas to achieve its ends.

4 Government expenditure is clearly (fairly) directly within the government's control and provided it does not see a Balanced Budget – that is an equality between current expenditure and tax revenue, as an end in itself – it can adapt its spending to the level it thinks appropriate for National Income determination. However, it is not a straightforward neutral and technical problem, for the government may spend more than it really desires for income-determining purposes because of political commitments to defence or social policy, or less than it really prefers because it cannot get the necessary parliamentary procedure completed quickly enough. This technique, too, tends to be long-winded.

Before we turn to the problems of the price level and to some implications of all this, some of general interest and some specific to the firm, it is worth drawing our arguments together and emphasizing some points.

We must not allow our earlier warning against mistaking an increase in employment for an equi-proportionate increase in output, or vice versa, to mislead us; the kind of factors which govern income also govern the level of the employment in the economy. If we wish to increase output, income, or employment then we must raise expenditure within the economy. It is for this reason that wage cuts are an ineffective means of coping with unemployment as they lower the expenditure levels and so lower incomes and employment in the short-run.

This is a convenient example of how hazardous it is to argue from micro-economics to macro-economics. (For instance, those workers who now have less spending power must unwittingly cause unemployment elsewhere.)

This brings us to an opportunity to make the normative/positive distinction. For many people the concept of expenditure carries moral implications; to the puritan it speaks of self-indulgence and to the profligate it simply offers opportunities. Now economists have as much right as anybody else to these normative or moral positions but they also have a further professional responsibility. And this is to recognize that spending and saving are not acts to be judged for themselves as though they were in a vacuum. It is a matter of positive economics that they have manifold and important effects far away from the individual spender or saver. *They are not economically neutral.*

It must always be remembered that rising expenditure can allay unemployment and increase output and income. However, as soon as we turn to the problem of the price level it becomes apparent that the expenditure level is a dangerous and difficult medium for government policy and we come to a 'choice' problem that is characteristic of economics.

How is the general price level determined? The outline of the answer is that it is determined by the expenditure level just as are the levels of output, employment, and income. The relationship of these last three with expenditure is fairly straightforward even if it is not a proportionate one. The relationship of expenditure to the price level is more elusive and varies with circumstances. Firstly, it is not the expenditure level itself that governs the price level but *the expenditure level related to the level of employment or resource use.*

The general form is that an increase in expenditure causes increases in output and practically no price changes in times of higher unemployment; nearer full employment, spending increases can lead to both price and output increases and at high employment levels expenditure increases cause price increases but practically no changes in output. If changing expenditure leads to price changes they are known as money sector effects and, if to production and employment changes, they are called real sector effects.

We can think of all this as a spectrum at one end of which expenditure changes have their effect mainly on the real sector and thus on resource use and employment, and then, as we move along the spectrum, the effects fall on both the real and money sectors until, at the other end, they are practically all on the money sector – the price level. At this stage an important reason why there is little effect on employment is simply that *there is already full employment*. Looked at another way, real sector increases in National Income are an increase in the real production of goods and services but a money sector National Income increase means simply that its money value has increased – it is simply the same real output at higher prices.

Two warnings. It must not be thought that there is some kind of trip-wire on one side of which expenditure increases benefit the economy and on the other damage it; the analogy of the spectrum is more helpful. Secondly, the term 'full employment' has no precise meaning; to some it is 1½ per cent of the working population unemployed, and to others 3 per cent or more. At both these levels we would expect to get money sector effects, that is inflation. The rate of inflation experienced with a given level of employment will vary from time to time, and from one country to another. There is no general rule that a given employment level will produce particular price behaviour

The general pattern we have described here is well substantiated in the real world. The countries which have had the slowest rates of inflation in recent years (Guatemala, Venezuela, Honduras, United States, and Luxembourg), have also had relatively high levels of unemployment. This is not to deny that sometimes a country can have both high employment and price stability in the short-run nor to say that there is a clear relationship with the most inflationary countries having the highest levels of employment. *That* is a different row of beans altogether; each particular inflation will have its own particular explanation and economists would turn to a number of factors according to the circumstance. We are simply drawing attention to a very well-established and widespread relationship, sometimes called the Phillips function after its progenitor, that holds that the inflation rate is zero at low employment levels and becomes higher the higher the employment level in a given economy.

This brings us now to a problem of choice that helps to clarify the role of the economist.

If a politician offers the twin goals of long-run full employment and price stability he is most likely engaged in a confidence trick and, possibly, self-delusion. The economist's attitude is that society has to choose between them or, put more realistically, that it must choose only partly to attain both goals and thus settle for a mix. It might be 3 per cent unemployment and 2 per cent inflation, or rather less unemployment but more inflation, and so on. The economist's job is to make the issue clear and then recommend policies consistent with whatever choice is made. There is, thus, no such thing as a 'correct' economic policy in the sense that Einstein was correct about relativity. The most we can hope for is that a mix of goals is clearly specified and that the government can then find policies consistent with them; economic policies can be correct, not in some moral or absolute sense, but only in that they are consistent with a stated group of aims. Particular tax and subsidy policies are consistent with re-distribution of income; certain tariff, quota, and currency policies with a Balance of Payments surplus, stable exchange rate and aid to underdeveloped countries; and specific exhortatory, physical, fiscal, and monetary policies with gaining the expenditure levels appropriate to the required mix of economic growth, employment, and GNP levels.

We say appropriate levels because in this sort of work there is no possibility of getting it perfectly right – to the nearest shilling, or the nearest £ million come to that. (A more detailed account of these policies is given in Chapter Five.)

We come now to some general conclusions. Firstly, there are no ready-made rules about the proper level of savings, investment, consumption, taxation, government spending, exports or imports. There is not even a general *economic* rule in favour of reducing taxation; each circumstance is different and each case needs to be judged separately. There may be, of course, personal or political reasons for lowering taxation, or changing any other policy.

Secondly, the well-tried favourite that all the UK's economic problems are due to the level of government expenditure gains no support from economists. Apart from the fact that it is just not true that the UK has proportionately high government expenditure[2] the important thing is that it is aggregate expenditure that is, or is not, inflationary and government expenditure is no more 'guilty' than any other part of the total spending. Thirdly, the Budget must not be seen as an exercise in book-keeping. The main purpose of taxation is not to meet the government's bills, for that would be to pretend that the Budget is economically neutral. The budget relationship between taxation and government spending is economically positive. The government is well aware of this and the main purpose is to get levels of taxation and government spending that suit its economic policy. It is concerned with keeping the accounts straight in the sense of being accurate, but it is not its intention to balance them for the sake of balancing them.

This brings us, fourthly, to the important distinction between economic and financial matters. Newspapers and television often talk of an economic crisis when they mean a financial one and vice versa. We must draw our lines a little more carefully.

Sometimes, government, health, education, or roads programmes are cut back or postponed on grounds of financial stringency. This does not mean that the government cannot pay its bills, it simply means that the government has chosen to save a particular proportion of the nation's resources and to allocate these elsewhere. The original proportion of the nation's resources was chosen by selecting a particular mix of goals, which led to a certain level of government spending and the allocation of this government spending between competing uses was done, one must hope, by ranking claims, and some claims were ranked last. That is, it was a matter of priorities and not of being unable to afford it. It is not necessarily true that the choice of goods leads to a certain level of government spending because this itself may be one of the goals, but it remains, nevertheless, a matter of priorities. *The community can easily 'afford the expense' but it is not always prepared to pay for the road, hospital, or school by forgoing the alternatives – which might be in either the public or private sector.*

The widespread confusion of economic with financial difficulties leads both to careless use of words and to policies which are distinctly muddled at the edges. Britain's international trade problems sometimes lead to phrases like 'economic crises' and 'international bankruptcy'. A glance at 'per capita' GNP figures for a hundred countries would show the real meaning of 'economic crisis' compared with Britain's position, and a further few minutes looking at the thousands of millions the UK can borrow compared with what the poorer countries can borrow, or at the UK long-run assets, soon shows who can get the money, who is credit-worthy, and who is not.

An example of the second point is that if, as part of an anti-inflationary policy, the government restricts government spending and chooses to limit hospital building then there is both less aggregate spending (all other things being equal) and it has also released construction resources which will mostly be employed elsewhere in the construction industry, in the private or public sector. In contrast, a tighter credit policy may also manage to lower aggregate spending but its effect on resource allocation will be much more diffuse.

A rather different point is that if, in the name of economic strategy, the local authorities are refused funds to finance mortgages for older houses it is difficult to see that this has any economic consequences worth talking about. The community has exactly the same number of old houses as before; the main effect achieved is on the exchange of money for houses between would-be owners and existing owners. This seems an unimportant result. This sort of anomaly results from confusing economic considerations with financial ones and, in particular, from thinking that financial rationing is the same as resource allocation.

Let us turn, in conclusion, to the specific implications of National Income analysis for firms. Clearly, the first point is that the level of economic activity modifies the position of the firm. High levels of economic activity make it easier, in general, to sell its products but harder to buy its inputs – raw materials, components, manpower and so on; harder in that it will have to compete for them more vigorously, both in efforts to find them and in prices to attract and hold them. The opposite is not quite true, for in times of unemployment, although labour may be easily available, it is not true that the price of labour (wages) is likely to fall.

Not only does the economic activity level help to describe the environment of the firm but it cannot be relied on to be stable so that the firm is constantly facing changing resources and changing objectives.

The level of economic activity varies because there is nothing to make it autonomously stable and, certainly, nothing to make it stable at a full employment level. So that even if the government did not intervene in the economy, firms would still have plenty of troubles: government policy simply modifies the form of the problem. Due to the multiplier and the accelerator, the investment goods industries are more unstable than the consumption goods industries and these firms must be particularly on their guard against a confident feeling of safety.

The case for government policy is based on two simple propositions: firstly, the economy will not automatically produce the goals of government economic policy so that government intervention follows, not for its own sake, but as a matter of consistency once the goals are chosen; secondly, government policy may be against the interest of individuals or firms but that is not to say they are against the public interest for the two do not run in tandem. People who seek to minimize the role of the government in the economy are, whether they know it or not, engaged in either an untenable position or a value judgement. It is untenable that furthering the interest of the individual is necessarily advancing those of society; alternatively, since to intervene 'correctly' is consistent with the goals, opponents of government policy can only really be disagreeing with the objectives. Now these are interesting matters of opinion that can be debated all night but, whatever view is taken, that view is a value judgement and nothing to do with positive economics.

Whatever our personal prejudices, we can take it that the government will be intervening in the economy for some time to come, mistakenly or otherwise. Firms are affected in two main ways.

Some government policies like physical, monetary, budgetary, and exhortatory techniques operate on the economic environment in what it takes to be the national interest. As their environment changes so do the ranges of tactics and stratagems from which the firms have to choose.

Other policies act in a more direct way on the actual decisions of the firm so that the firm becomes more likely to make a particular choice from the new menu of choices before it. A few examples will suffice for the moment. Regional policy weights the choice of locations open to a firm in a particular fashion and monopoly policy favours some courses of action, rather than others, of all those technically open to a monopolistic firm. Lastly, if society does not like the firm's behaviour or, perhaps, output or employment levels, even within the modified environment and with particular choices made to look more advantageous than others, then it can take over ownership or at least provide partial finance. Examples of partial finance are government contracts to aircraft companies, at least partly intended to keep up employment, tax allowances for particular purposes like research, and financial advances from bodies like the Industrial and Commercial Finance Corporation and the Finance Corporation for Industry which get some of their money from the government-owned Bank of England.

The most obvious form of government ownership is the nationalized industries, but the local authorities, the National Health Service, and bodies like the National Research and Development Corporation also come within the public sector.

There are very grave problems indeed in public sector economics, reserved for a later chapter, but our concern is with the way the level of economic activity affects the firms in the private sector, that is, with controlling the environment and with manoeuvring firms to act in a particular fashion out of all the options open to them. We have now outlined, in a fairly intuitive fashion, the factors governing economic activity. We will turn to the problems of policy in Chapters Four and Five.

References

1 Keynes, J. M., *General Theory of Employment, Interest and Money*, Macmillan, 1936.
2 See Chapter Six.

Bibliography

Bray, J., *Decision in Government*, Gollancz, 1970.
Pen, J., *Modern Economics*, Penguin, 1965.
Schultze, C., *National Income Analysis*, Prentice-Hall, 1967.
Stewart, M., *Keynes and After*, Penguin, 1967.

CHAPTER FOUR

How the Economy is Managed: Economic Policy 1

It is not easy to write for a non-specialist reader about government economic policy. The economist knows he is dealing with a changing, complex set of mutually dependent variables, where little has been tested and less proved, while the manager-reader, in contrast, tends to look for certainties. However, he is not looking for just any certainties for he wants those which are consistent with his own prejudices and convenient for his own firm. He wants to hear that tax rates are a disincentive and that the nation's economic ills can be laid at doors other than his own and attributed to that single cause he has himself long blamed. The economist fails him – but is true to economics – by neither proving nor disproving any disincentive effect in taxation and by seeing the economy as very complicated with no single factor to explain one of the economy's difficulties let alone them all. No serious analysis of the UK economy points to just one factor, attention to which will resolve everything. As we have said before, the economy must not be thought of as a car with a simple faulty component the replacement of which will lead to a smooth performance.

When laymen are so sure they know the answer and economists so sure that nobody does, it is not surprising that economists are seen as experts who fail to deliver the goods, and economists, in their turn, see the lay public as Philistines.

It is even possible that both parties are right. Economists *have* got very little in the way of more than handy guidelines for economic policy and laymen *do* reveal a combination of ignorance, innocence, and intellectual bravado when considering macro-economic problems. The mutual disappointment between laymen and economists is due to the former asking the latter to do the impossible and the latter's irritation that their relative impotence is on public view. The general air of suspicion is not helped by some economists taking an over-confident and bold line on complex problems and some laymen wilfully ignoring explicitly stated qualifications to the argument.

The role of the economist can be succinctly stated. It is to help formulate thinking about economic policy by separating the positive from the normative, by identifying inconsistencies in policy aims and by quantifying the alternatives where possible.

Let us briefly take each in turn. *Separating the positive from the normative* means no more than distinguishing opinions from facts. In economics, this commonly helps to sort out means from ends. Many policy disagreements turn on opinions being treated as facts – not open to discussion – and disagreements which are apparently about means, are really about ends.

A few examples will suffice. The layman is apt to treat stable exchange rates as a policy goal so self-evident that it is beyond discussion. It is not too difficult to see it as a constraint; exchange-rate variations are a policy not open to us and, in consequence, we accept lower economic growth rates or employment levels than we would otherwise. What needs emphasis is that it is a normatively chosen constraint – *we do not do it because we do not want to*; this is quite different from a positive constraint in natural science – the chemist does not turn base metal into gold because he cannot.

A further example of the positive/normative confusion in the public mind can be found in the subject of taxation. Many people who dispute the form of taxation are really discussing the purposes or, at least, should be; different forms of taxation are often discussed as though they were different ways of doing the same thing. Purchase tax and income tax cannot be sensibly debated as simply alternative ways of raising a given sum of money. They are ways of doing two quite different things. To see them as alternative techniques for raising a given sum is to mistakenly see the purpose of taxation as that of balancing the books as though it were part of society's housekeeping.

In emphasizing the positive and the normative, the economist helps to get the nature of the policy objectives and constraints recognized. The objectives are always normatively chosen but what is hoped is that they can be knowledgeably chosen. The constraints are likely to be either resource constraints – we cannot build all we would like at the same time – or normative policy constraints in that the realization of one objective acts as a constraint by partially or totally eliminating another objective.

This brings us to the second role of the economist – the *identification of inconsistent policy objectives*. It also brings us back to our earlier emphasis on *choosing techniques to suit maximands selected on a normative basis from the constrained choices open to us*. All this means is that we must choose as best we can from the list. For instance, in our later discussion of nationalized industries we shall see that the dispute about the two investment criteria, of internal rate of return and present value, is really about objectives and not about means. Once the objective is specified then only one of the criteria is consistent; we choose between means by choosing between ends.

We shall see that the problem of identifying inconsistent policy objectives is largely the problem of discovering the exact nature of the policy constraints at a given time. The difficulties facing the economist are brought into focus by the following examples.

The economist meets his first difficulty in *scientifically* demonstrating causal relationships between one factor and another. Plausible relationships are easily found – between, for instance, the birth rate and the number of storks – but proof is another matter again. A second difficulty is that, even if a relationship is found, it is most unlikely that it is an economic 'law' that holds good for all circumstances. If, for instance, a connexion is found between economic growth and a Balance of Payments surplus it is most unlikely that it is applicable on a worldwide basis. All countries can experience growth simultaneously but, by definition, they cannot all experience a surplus at the same time; if some are in surplus then others must be in deficit. (This assumes the same Balance of Payments accounting conventions throughout the world.) Thirdly, when a relationship is set up it is another matter to know the form of the relationship – that is, what magnitude of payments surplus is associated with any given growth rate. Such information would be very operational; the idea that some more economic growth would cause a trade change is no more than mildly useful.

Fourthly, even if a quantitative value could be put on the relationship it could not be expected that this value would hold in other economies or in the particular economy at other times. Once the structure of the economy had altered – and maximands and resources are always in flux – and once any of the parameters have changed, then so will the specific relationship earlier identified. A given growth rate may stimulate a particular trade pattern but it cannot be uniquely identified with that pattern, for if the exchange rate were to alter then a different export/import ratio would be associated with that growth rate. It was for this reason that, by and large, in the 1960s Western Germany could have fast growth and a payments surplus, but the UK found further growth associated with payments deficits.

We can represent all this diagrammatically in Figure 1.

The choice frontier might be in three possible quadrants and society has to (normatively) choose one of the combinations open to it, either on F_1F, high growth and payments deficit, or on FF, some combination of growth or surplus, or on FF_2 – high surplus and a falling living standard. Once the alternatives are reliably identified, a combination such as A might be chosen in which some growth is sacrificed for some surplus, or vice versa. It is feasible that quite soon the choice frontier will shift and the growth rate G now becomes associated with the payments position at A_1 or A_2. We can never know what we are choosing between, and, therefore, what we are doing, more than temporarily.

FIGURE ONE

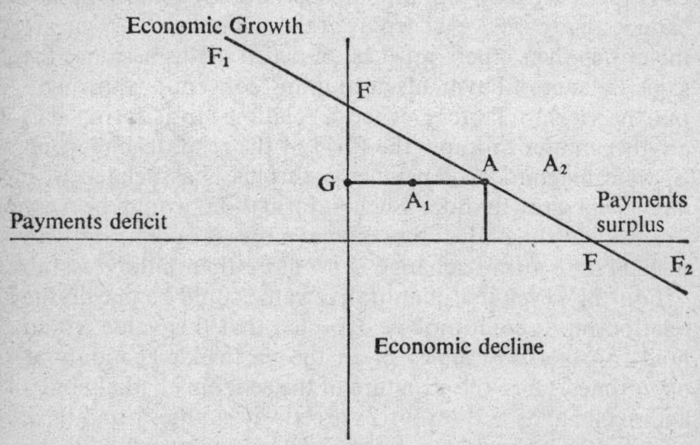

A catalogue of the goals of government policy soon shows that the real world is more complex than our diagram, for the government chooses not between two goals, but between five or six according to circumstance. The government will not have to choose between as many alternatives as it has objectives, for some of the objectives will be complementary to one another rather than inconsistent. More of one will mean more of the other like CC in Figure 2, where axes Y and X might represent Balance of Payments surpluses and stable exchange rates.

These are probably related in this way, at least in the short-run, for the greater the current surplus the easier it becomes to stabilize the currency exchange rate against devaluation. However, the experience of Western Germany in 1969 has shown that chronic long-run surpluses lead not to stability but to an upwards revaluation.

Yet other government policy goals may be totally unrelated to one another like C_1 in Figure 2 in which higher and higher values of Y (employment level) are consistent with the same value of X (aid to underdeveloped countries) which have no effect whatsoever on one another.

FIGURE TWO

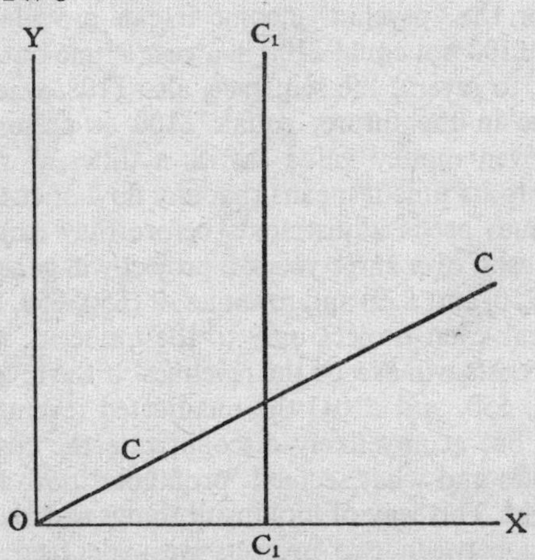

The economist's task is to identify which of the economy's objectives are related and in what way. Aid policies are somewhat separate but economic growth, price stability, full employment, stable exchange rates, payments surplus, and redistribution of income are all interrelated.

Having seen something of the normative/positive distinction and the problem of identifying inconsistencies, let us now turn to the question of *quantification*. We will discuss two examples – those of discounted cash flow, and cost-benefit analysis – as they both also bring us to relate means to ends and to seek consistency between the two.

DCF is a technique designed to accommodate the difficulty already noted in Chapter One that money takes on a different real burden, if it is a cost, and a different real benefit, if it is a revenue, according to the time at which it occurs. The result is that £100 does not equal £100 in a year's time but, perhaps, £108. The 'value' which we put on a sum of money varies according to time, partly because we could earn interest on it if we gain it earlier, and partly because, mainly due to uncertainty, we prefer to have the money now rather than later. The earlier we have a given sum of money the greater the 'psychic' income it can provide. So, not only does £100 not equal £100 in a year's time but, perhaps, £108, but, to reverse the sequence, also £108, whether cost or revenue in the future, equals £100 at present values. Since a given money value carries a different real value according to its time it means that any flow of cash in costs and revenues needs adjustments before they can be compared. Thus, if over three years, a project will generate costs of (£100, £50, and £50) and revenues of (£50, £50, and £100) it is a break-even project only to the innocent. When adjusted the costs will exceed the revenues. If the revenues had been (£50, £50, and £101) the unadjusted revenues exceed the costs, but at any likely discount rate the costs exceed the revenues and what seemed 'profitable' now appears in its true light. This way of looking at things can also usefully distinguish between two projects with identical costs but different time profiles in their revenues like, (£100, £100, and £50) and (£50, £100, and £100). Two such projects are identical only in their simple revenue totals but not in their revenue time profits nor, indeed, in their discounted revenue totals.

So far, so good. Two problems remain in the choice of the discount rate and the discounting techniques. The discount rate, 8 per cent, 9 per cent, or whatever, can only be normatively chosen. Those who would receive £100 today for £108 in a year's time have a discount rate of 8 per cent and those for whom it would have to be £110 have a 10 per cent discount rate and so on. (This is not mathematically quite exact since 8 per cent of £108 is not £8 but it will do for present purposes.) The point is that there is no scientific way of choosing a discount rate; every person must make up his own mind about how much he prefers jam today to jam tomorrow.

Still, so far so good. The difficulty of the appropriate discount technique needs a careful approach. There are two main rival discounting methods. The *present value approach* chooses all projects where the future revenues, adjusted for futurity, exceed future adjusted costs. *This allows the investor to choose that discount rate which corresponds to his feelings about the present and future.* On the other hand, the *internal rate of return approach* chooses all projects where the discount rate that equates future revenues and costs exceeds the appropriate interest rate – presumably that is the current interest rate, in most cases.

There has been acrimonious debate in the trade on which of these techniques is the 'better'. Such discussions are misplaced for one technique suits one objective, and the other a different objective – and yet other techniques suit yet other goals.

The internal rate of return correctly chooses those projects which will yield the greatest stock of assets in the future but without preference given to one time rather than another. Thus it is appropriate for the Stalinist maximand in the USSR of the 1920s and 1930s, for the Russian government chose projects which yielded the greatest future aspects without reference to present costs (forgone present consumption), or the time at which the assets become available. It is also suitable for a firm seeking growth irrespective of forgoing present profits or the time at which it gains the desired size. Firstly, to choose a bizarre but convincing example, it is appropriate for an Egyptian Pharaoh to ignore council hut building for his subjects in order to build pyramids that generate foreign exchange earnings 4,000 years later, enjoyed by a quite different society.

The present value approach, however, is appropriate for those firms and societies that do care about the time profile of consumption and it enables them to use a rate which reflects their own preferences about jam today against (more) jam tomorrow. It is thus suitable for societies that want a mix of consumption today and more consumption tomorrow other than the mix characterized by the minimum possible consumption today and the maximum possible stock of assets at some unspecified time in the future. The present value technique is suitable for those firms who have the currently fashionable objective of maximizing net present worth.

This seeking of consistency between ends and means is characteristic of the economist's mode of thinking. Another facet is that there are objectives which suit neither the internal rate of return nor the present value techniques. It is possible to visualize a firm that sets its objective as eliminating a deficit in the shortest possible time – the 'Beeching maximand'. If the deficit were 9 and alternative projects offered 10 in one case and then nothing, (A); 5,5, and 1 in a second case, (B); and 5,5, and 10 in a third, (C), neither of our previous techniques is consistent with the goal. Clearly, project A must be chosen for it eliminates the deficit quickest. It was following out the consequences of the Beeching plan for British Rail in this sort of way that brought it into disrepute since it implied an unusual choice in this sort of case. It also ignored social costs and benefits.

We can see also this consistency between ends and means in cost-benefit analysis, the second of our quantitative techniques. *Cost-Benefit Analysis* is quite inappropriate for commercial companies although this fashionable term is sometimes used as the equivalent of investment appraisal. In fact, properly used, cost-benefit analysis seeks to overcome the reliance on 'profits' in *public enterprise projects* as a criterion for comparing the efficiency of different industries or for ranking alternative investment projects. This is because money prices do not measure true benefits and money costs do not measure true costs. Cost-benefit analysis seeks to juxtapose true benefits and true costs in order to appraise alternative projects.

It adjusts money prices in seven different ways in order to approach more really to the truth. Externalities are taken into account to correctly gauge costs and benefits; thus if an airport necessitates soundproofing in buildings unconnected with the airport, then this is a true cost, whether or not the airport authority is legally responsible for the expenditure. If an irrigation project lowers dredging costs elsewhere on the river, then that is an external benefit. Costs and benefits are also adjusted for 'rental' effects: if workers on a motorway can earn £50 per week and their next best alternative is £30 per week on council building sites then the true cost to society is not £50 at all but £30 of alternative production forgone. Similarly, if a motorway raises the value of existing garages at entry points the community is neither richer nor poorer for it has no more assets than earlier and such effects must be ignored. Taxes and subsidies must be allowed for, too; if a motorway saves 1,000 gallons of petrol a day the true benefit is not measured by 1,000 times the market price of petrol but by 1,000 times the real cost which will be less of tax; the reverse argument applies to subsidies. The fourth method by which we seek out the so-called 'shadow prices',[1] instead of the market money prices, is by eliminating monopoly elements in prices. Our last three adjustments are more speculative than precise; we acknowledge the existence of a multiplier in the case of government expenditure so that the final effect is measured, rather than the immediate effect; we try to evaluate collective goals which have no price to users, like bridges and street lighting, and we try to put a value on intangibles like the effects on scenery.

Whether or not these techniques should be used is simply a matter of the goals of society. Those who think that money prices are reliable will use them to analyse social investment while, on the other hand, those who are seeking, not to balance the books, but to balance true costs against true benefits will find cost-benefit analysis consistent with their objective. The implications of all this should be made clear; the cost-benefit analyst has no interest in whether a nationalized industry makes a profit.

Having looked at the economist's work in separating the positive from the normative, in identifying maximands and constraints and in quantifying where possible, let us look at the *raison d'être* for government policy and its main forms.

Once there is a list of goals of government policy then government intervention in the economy necessarily follows unless, of course, the sole item on the list is that the government should not interfere. Government intervention follows because we want a particular allocation of resources that a free price mechanism cannot produce. We cannot rely on money prices to bring it about any more than we can rely on the various forms of expenditure to autonomously give us price stability, economic growth, full employment, and so on.

There are a number of different ways of classifying government economic policy. No one classification is logically superior to any other any more than numbering all the houses on one side of a street as even numbers is any more sensible than giving them odd numbers. A classificatory system is a sort of mental filing cabinet and all we can ask is that it is convenient; we must expect it to be more convenient for some purposes than others. We will look briefly at two classificatory systems and then in detail at a third

It is quite customary to categorize government policy in the following way: *fiscal*, or budgetary policy, is the use of taxes and subsidies to produce both levels of total spending and patterns of spending different from what would otherwise occur; *monetary* policy changes the volume of money and the level of interest rates for the same purpose; *physical* policy is the use of legal forms like fixed prices, rationing, and licences, while *exhortatory* policy is the use of propaganda to induce people to act in some particular way, savings campaigns and National Productivity Year are examples. We can summarize all this by saying that if a government wishes fewer foreign cars to be imported then it may raise their prices (fiscal), raise credit rates (monetary), ask people not to import (exhortatory), or simply make it illegal (physical). A final form of government policy is *Prices and Incomes Policy* which tries to influence prices and income changes to combat trade deficits and inflation and to promote economic growth. It does not fit easily into our classification since some of its techniques, like Statements of Intent, are really exhortatory and others, like Prices and Incomes Acts, are really examples of physical policy.

This classification is traditional but it carries two difficulties. Firstly, it does not rely on a single identifying characteristic; one of its criteria is the arm of the government that executes the policy – monetary policy is in the hands of the Bank of England and fiscal policy in those of the Treasury, while in the case of physical policy the identifying characteristic is the nature of the policy itself rather than the government department responsible for its operation. A second difficulty is that this classificatory system does not relate the categories to the policy objectives. A categorizing policy that takes no account of the purposes of the material classified is unlikely to be especially useful. It is rather like giving buses route numbers and destination notices without relevance to where they are actually going.

A second classification turns on what the government does. This provides five groupings: it exhorts, it manages the National Debt, it provides financial assistance, it regulates in two quite different ways, and fifthly, it provides goods and services in three quite different ways.

It is helpful to note three ways in which these policies operate. Firstly, they change the options open to the firm, secondly, they may favour one option rather than another and, finally, the State may undertake a particular policy where no firm chooses the policy option judged to be in the public interest.

Exhortatory policy has already been noted and National Debt management is self-explanatory.

What are the two ways in which the government regulates the economy? Firstly, it provides legislative constraints which eliminate policies which are otherwise technically possible. Factory Acts rule out particular working practices, Town and Country Planning Acts illegalize certain land uses and so on. Other regulatory devices allow certain courses of action, but make them less attractive by making them more expensive or vice versa. Taxes, subsidies, and credit policies fall in this category. We have already seen something about the effect of taxation and subsidies in their macro-economic role of withdrawals and injections which help to determine the level of aggregate demand within the economy. Let us look briefly at indirect taxes, such as purchase tax and excise tax in their micro-economic context.

There are two aspects: what is the effect on prices and on resource use?

It is common to attribute higher prices to tax changes but the matter is not straightforward. When taxes fall on goods the price of which can be raised with a minimal effect on demand, that is when demand is largely independent of price over a plausible price range, then a tax increase can be expected to lead to a price rise of the same order of magnitude. (Put technically, price increases equal, or approximate to, tax increases when the good is price inelastic.) The tax increase thus falls on the consumer. In such cases, tax increases are not likely to produce any serious shift in output and so resource use is virtually unchanged. Such tax policies do not constrain the firm much, though they must constrain the customers into spending less elsewhere. Usually this switching effect is so general that its effect on any given industry is negligible.

The rest now clicks into place. If the demand for a good is very dependent on price (elastic demand), then prices do not rise much but output falls and so the tax burden falls mainly on the producer and there is a reduction in sales. Vice versa will hold for subsidies – which are simply 'negative taxes'.

Clearly many goods in the real world do not fall at either end of the spectrum we have postulated and the tax burden is roughly shared in such cases. A casual experience of the real world confirms this analysis, which suggests that there is no general 'law' and that tax effects vary from case to case.

The 1969 UK Budget provided a convenient example. A tax was levied on cat food; this produced an immediate increase in the price of almost all tins of cat food in a general attempt to pass on the tax increase. Within a few months different brands had met a different falling off in sales and so the tax was not eventually passed on equally in all cases, as the prices had to be adjusted in some cases.

Unfortunately, it is not obvious to the public that not all taxes are passed on because so many taxes in the UK *are* passed on, for British taxes tend to fall on tobacco, alcohol, and petrol which happen to be large company industries characterized by a lack of price competition. In these oligopolistic industries, companies tend to charge the same prices and compete by techniques such as advertising.

A further example will illustrate the applicability of our approach that the effect depends upon the responsiveness of demand. Clearly, if demand is responsive to price then a tax increase means fewer sales and so a small tax revenue. It is this which explains some aspects of the British tax structure; the major aim is often to get a significant tax revenue either in a housekeeping approach to the Budget or in a more sophisticated approach that sees the tax revenue as a withdrawal. In either case, the way to get tax revenue is to look for goods the demand for which is price-inelastic – and so it is no accident that taxes fall repeatedly on tobacco, alcohol, and petrol.

Not only is it not an accident, it is to be seen as consistent with the objective of raising tax revenue simply rather than as a Puritan Plot against smokers, drinkers, and drivers. More of this in Chapter Six.

In these ways the government weights the attractiveness of different ways of spending money by changing prices and sometimes it may tax or subsidize for this motive alone rather than to gain revenue.

Let us turn now to the provision of resources on a total or partial basis by the State. On a partial basis, it may provide finance to meet a particular need without assuming day-to-day responsibilities. Regional employment premiums and tax allowances for investment are well-established examples.

If, in contrast, the State takes over the complete production of a good or service it has an immediate problem of choosing from three pricing policies. These remarks apply not only to obvious cases like the nationalized industries but also the British Museum, public parks, and municipal swimming pools.

Firstly, some services, like the police, the libraries, and the roads, are financed out of general taxation and then provided 'free'. The economist regards very little as free and certainly not these services. A zero price is as much a distortion as any other price that fails to measure true cost – in this case the opportunities forgone when resources are allocated in this way. In these circumstances the 'price' paid by an individual depends upon his tax burden and so (very approximately) to his ability to pay. In no sense is what he pays related to the use he makes of the product.

Secondly, the activity may be financed like any other commercial activity such as, say, the nationalized industries. Then the price paid is roughly related to the use made, in the sense that the more you use the more you pay. Thirdly, the production of goods or services may be financed with reference neither to ability to pay nor to the use made. Examples are the National Health Service and the BBC, both financed from specific forms of revenue which are the same for any given class of person, independently of how often he is ill or how frequently he listens or views BBC programmes.

It should by now be clear that there is no consistency whatsoever in the way in which public services are financed.

At first look this may seem anomalous, but, in fact, to expect a pattern in these matters is to prejudge the issue; firstly, it mistakenly presumes that there is a known 'correct' way to finance public enterprises and, secondly, such an attitude misses the point that we look for *pricing policies consistent with objectives.* Since different parts of the public sector have different objectives we must expect, rather than abuse, the existence of different pricing policies.

This second classification has the advantage that it brings

out clearly the difference between policies which eliminate particular options, otherwise open to the firm, and those policies which do not rule out particular courses of action, but weigh simply some against others. However, there is no escape from the *pot-pourri* nature of this scheme since, for instance, fiscal policy comes in both categories.

The third categorization is more faithful to the nature of economics in that it relates policies clearly to the goals.

Employment Policy tries to get values for C, I, and G appropriate to full employment. Taxes, subsidies, interest rates and direct government spending can all hope to have this effect, as can physical restrictions.

Anti-Inflationary Policy seeks values to those variables which produce full employment without generating inflation.

Economic Growth seeks tax, subsidy, interest rate, and legislative techniques which foster variables thought to be critical to the growth rate. Innovation, research, labour mobility, training, and investment are suggested candidates.

Balance of Payments Policies use tariffs, quotas, and credit policies to stimulate the required trade patterns.

Exchange-Rate Policies use legal sanctions to maintain particular rates.

Income Redistribution and Overseas Aid Policies use taxes and distributions to meet these ends.

We have already seen that monetary and fiscal policies are hampered by the difficulty of knowing what values to give taxes or interest rates to obtain the required level of expenditure. They also tend to be ineffective in the short-run. What else can usefully be said about fiscal and monetary policy?

A brief history would show that in the immediate post-war era, fiscal policy was given pride of place in the management of the economy but that in the early 1950s monetary policy staged a comeback in popularity. The achievements of neither era lead to any great satisfaction and this – coupled with advances in economic theory – have produced a position in the 1960s in which the techniques are used together rather than as alternatives.

In recent years monetary policy has been an awkward technique to use effectively, because, if interest rates are raised in order to discourage spending and combat inflation, the higher interest rates attract international money ('hot' money) to the UK which misleadingly enhances the Balance of Payments – which is difficult enough to interpret at the best of times. Conversely, if interest rates are lowered to encourage output and employment levels, money leaves the country to seek higher interest rates elsewhere and the Payments position is threatened. In other words, if monetary policy is to be effective within the domestic economy it tends to bring unwanted international consequences: it is a choice problem characteristic of economics.

The supply of money has also given the government difficulty. The government has been under pressure from the IMF to lay greater stress on the money supply as a control device. Simultaneously, some economists (the Chicago School) have advanced the position that the money supply is of crucial importance, but these theoretical analyses do not go unchallenged among rival economists and the present position is unconsolidated.

Whatever the validity of the theoretical position, there are a number of practical difficulties which greatly constrain the style of the government's monetary policy. The orthodox technique for altering the volume of money used to be open-market operations; this meant that the government changed the volume of money in the hands of the public by the purchase or sale of government bonds on the Stock Exchange. The immediate difficulty is that this alters bond prices, and so yields, and thus one group within the community bears the main brunt of a policy which is for the collective benefit as well as their own.

A further difficulty is that the government, in controlling the supply of money, must look to the volume of credit as well as that of cash. Recently, attempts have been made to fix 'ceilings' on the total level of bank advances. This technique can simply lead to forestalling in the medieval sense; companies negotiate overdrafts greater than they currently need in order to preserve their future position in the event of a new ceiling. In any event, such credit restrictions fall unequally on different firms according to their debtor position; there is no guarantee that the firms which do restrain spending are those engaged in frivolous production and those which continue spending are vital to the nation's economy.

Further difficulties arise in that a significant part of spending is under hire-purchase contracts. This spending is responsive to changes in the HP regulations, but the difficulty arises again that the adjustment falls on one sector of the economy – in this case the consumer durables industry and its HP customers. Another point is that ways are continually devised of out-manoeuvring the policy by devising hire-purchase contracts that fall outside the scope of the regulations – 'revolving' credit provides a convenient example.

This links on to a further point that, as the government learns to control the money supply, then further forms of money are invented – so-called 'near money'. The moment the government gets bank credit under control there is a boom in second mortgages and so on.

At times the government must feel that it does not have enough fingers to plug all the holes in the dyke.

Fiscal policy, as we now know it, started with the Kingsley Wood Budget of 1941. From that time, taxes and subsidies have been explicitly regarded as techniques for manipulating the level of aggregate spending rather than as an exercise in book-keeping.

One of the key concepts in fiscal policy is that of the automatic stabilizer; this is the idea that, for instance, as money incomes grow, any potential inflation will be eliminated as the higher spending power is diverted from the public by taxes. This hope has been vitiated in Britain because British taxes are so unprogressive.[2] In general, fiscal policy has been more effective at maintaining employment than output.[3]

It might seem that government spending can be easily controlled, but this is something of an illusion. Apart from the difficulties we have already mentioned, one source of trouble is that government expenditure is in the hands of a multifarious collection of agencies – the Treasury, all the ministries, the Universities Grants Committee, Nationalized Industries, ICFC; foreign governments with whom the UK has contractual agreements, and hundreds of local government bodies – each one of which feels that its programme should not be cut. In addition, few of them have spending programmes that can be modified easily and quickly for it is not sensible to leave the Forth Bridge nine-tenths finished.

One feature of the UK system is that the government can alter the tax rates from the day of the Budget announcements; a more common practice elsewhere is for the legislature to debate the tax change proposals before they become effective. This must lead to strange patterns of anticipatory spending between the proposals and the legislation.

The British system also has the advantage of the 'regulator' which is the ability to alter certain purchase tax rates between budgets as occasion demands.

Fiscal policy is a useful government technique which is certainly not sufficient to do the job by itself, but is a necessary part of the tools of economic management. Techniques of physical policy, like rationing and price control, are always used in war time and other crisis periods. This suggests that properly administered they can be effective, but that they are only used reluctantly.

Why are they unpopular despite the fact they work? First, they tend to need an elaborate administrative superstructure and so are resource-expensive – involving a high opportunity cost. But the main objection is that they involve close State intervention and many people resent this; it is, however, a normative aspect of the problem to which the economist can bring no expertise.

Exhortatory policy, like any other advertising, varies in effectiveness from case to case. There is no clear evidence that productivity slogans, savings campaigns and so on have any measurable effect. The most that can be expected is that such techniques help to formulate general public attitudes.

Two points to note are that exhortatory policy is most likely to be effective when the consequences to the individual of compliance are preferable to those of non-compliance and, also, that such policies contain an odd moral feature in that non-compliance can gain the collective benefit just as much as compliance. A savings campaign can produce an anti-inflationary effect, for instance, and so there is a collective benefit which is enjoyed by the non-savers who *also* gain the private benefits of expenditure.

Prices and Incomes Policy uses both physical and exhortatory techniques. Prices and Incomes Acts are examples of physical policy, and the work of the PIB is exhortatory; it is not so resource-costly as physical policy and is less highly charged politically but is probably less effective. It is a typical choice problem; more effectiveness means higher costs. Whether the present policy is an acceptable mix is a matter of opinion; no doubt society's preferences will change in this matter over time.

It is worth now giving some special attention to nationalized industries. No general judgement can be passed about them; we simply have to ask how far each one meets the current objectives – low prices (CEGB), larger scale of operation than possible under private enterprise (BR), better labour relations (NCB), or whatever other objective.

The economist is better employed addressing himself to the problem of pricing and investment.

Most UK nationalization statutes required the nationalized industry to 'break even taking one year with another'. This seems an eccentric goal for it carries the mistaken implication that a 'revenue to money costs' equality is the same as real benefits equalling real costs. This is just not true. However, it does coincide with the widespread and misplaced faith in the reliability of money as a measuring method and of profits as a measure of efficiency. This goal is thus likely to be popular and if it is chosen, for normative reasons, then the appropriate policy is to set prices equal to average costs, since then the operation will break even. But that is all we can be confident about. For some people it is enough.

A White Paper on Nationalized Industries in 1961 introduced the concept of financial targets, or rates of return, to be earned on capital. This, too, is eccentric both in relying on money prices and in using the return on capital as a cypher for the return on all resources. This is common practice in the private sector, but we might hope that the public sector could set an example in looking to the expertise with which all resources, like labour and land, are used. It is a curiously distorted view of the economic process to exclusively look to return on capital as a decision rule. A further difficulty is that an explicit rate of return on capital is surprisingly ambiguous when it comes to pricing policy and is, in fact, little help. A wide number of pricing policies may yield the same return on capital; one can easily imagine peak-hour travellers paying more, and off-peak customers less, and the new price pattern giving the same return on capital.

What else could we ask of a public sector pricing policy other than it should break even? It might be hoped that it would set prices that really measure true costs – and this means putting prices equal to marginal costs. This was favoured by a 1967 White Paper.

There is a political disadvantage in that this can mean setting prices so that accounting 'losses' are made in those industries where marginal cost is below average cost. This is the case where a further passenger can be carried, on a half-empty train for instance, at negligible cost – in such cases there are no extra running costs or capital costs; simply those of printing the ticket. This is no problem to the economist since to him the low price measures the low cost of production for that customer perfectly well. It is, however, an awkward point in a society not willing, for normative reasons, to accept loss-making activities. A further difficulty is that, in such cases, marginal costing may be inconsistent with simultaneously earning a return on capital.

However, the White Paper confused the matter further by allowing four exceptions to the marginal costing rule.

It was 'impracticable' to cost some services separately (railway left luggage offices). There would be 'good commercial reasons' in other cases (restaurants and hotels pay higher gas prices than domestic consumers because their demands do not vary much with price). It might be said in passing that it can be difficult to reconcile 'good commercial reasons' with the kind of objective often specified for nationalized industries. If commercial practices were what were required they could be left in commercial hands.

The third exception was where there were peak-load fluctuations in demand and the fourth was where a pricing policy could call forth extra business that would not require proportionately greater costs. (Off-peak electricity tariffs can lead to a greater demand that can be met with little change in operating costs.) Both these exceptions have the feature that they lead to fuller use of existing plant and existing assets.

These four exceptions are so wide-ranging that they lead to a suspicion that marginal costing is more honoured in the breach than in observance. However, it remains true that marginal costing is the pricing policy consistent with the objective of finding prices that measure true costs.

In 1968 a Treasury Memorandum to the House of Commons Select Committee on Nationalized Industries persevered with both marginal cost pricing and financial targets, but added the Discounted Cash-Flow principle to investment appraisal in the public services. This has the merit of recognizing the truth that money revenues and costs take on a different value according to when they occur.

Nevertheless the nationalized industries in general are a topic that gives economists more despair than delight.

References

1 This term is further elaborated in Chapter Seven.
2 See Chapter Six.
3 Dow, J. C. R., *op. cit.*

Bibliography

Caves, R. E., and Associates, *Britain's Economic Prospects*, Brookings Institution, 1969.

Denison, E. F., assisted by Poullier, J-P., *Why Growth Rates Differ: Post-war Experience in Nine Western Countries*, Brookings Institution, 1967.

Dow, J. C. R., *The Management of the British Economy*, 1945–60, Cambridge University Press, 1964.

'A White Paper on the Financial and Economic Objectives of the Nationalized Industries', HMSO, 1961.

'A White Paper on the Economic and Financial Objectives of the Nationalized Industries', HMSO, 1967.

CHAPTER FIVE

How the Economy is Managed: Economic Policy 2

This chapter is a continuation of the last, in that it deals with government policies which involve intervention in the economy. Now we are concerned with policies that affect the structure of industry. Our purpose is to analyse why governments seek to change the structure of industry, how they have done this and what have been the results. We are concerned with change, not with describing the present structure or how it has come about. The latter are outside the scope of this book. Our concern is solely with government-induced change. We will deal with techniques which are general or designed to affect the firm; we exclude policies which take as their unit an industry, *eg*, subsidies to agriculture or special financial aid to the shipbuilding industry. This exclusion is necessary for reasons of space but does mean that some interesting effects will be ignored. For example, shipbuilding receives considerable subsidies here, in other European countries, and even in Japan; the result of this worldwide subsidization is a much greater output of ships than would otherwise occur. Thus, indirectly, the shipping industry and world trade is larger than would occur under free market conditions.

Why do governments intervene in the structure of industry? In general terms, such action rests on a belief that, left to itself, the private enterprise system does not produce the best possible results in terms of the objectives outlined in Chapter One.

We will deal with three aspects of this general problem *viz*:

1 Changing the structure of those industries where it is held that the existing number of firms, size of firm and inter-firm relationships could be improved in terms of promoting the 'public interest'.

The geographical distribution of firms may be seen as an example of this, with which we have already dealt in Chapter Two.

2 Reducing rigidities in the economic environment by facilitating the mobility of factors of production and improving their quality.

3 Promoting and shaping technical change.

1 *Changing the Existing Industrial Structure*

Unlike in the USA, dislike of a particular type of market structure (*eg*, a monopoly) in the UK has not been based on disquiet about the concentrations of economic and social power which they have brought about. This is generally true; however, dislike of concentrations of power was probably important in the debate which led to the 1948 Monopolies Act, and the 1965 Monopolies and Mergers Act contained provisions controlling further concentrations of power in newspaper ownership, largely a reaction to the growth of the 'Thomson Empire'. Recent popular discussions about the effects of mergers have started to emphasize the 'political' results of concentrations of power although this has not been a consideration in the official intervention dealt with later. Unlike in the USA, however, we do not frequently produce statistics to show, *eg*, the proportion of industrial output controlled by the 200 largest firms or a few influential families.

British legislation is not concerned with structure as such but with the performance resulting from a particular structure. The performance is assessed in terms of the 'public interest' which, as we shall see, has proved difficult to identify in practice, but is often expressed in terms of output (its price and quantity) and efficiency (in terms of the conversion of inputs into outputs). This concern with output is another distinguishing characteristic between this policy and types 2 and 3 – which are concerned with inputs.

What is the association between market structure and performance? Structure is normally considered in terms of a contrast between monopolized and competitive industries. Monopolies are popularly supposed to charge higher prices, restrict output, earn high profits, and restrict consumers' choice by reducing the variety of goods available and suppressing new products which might harm their profitability.

There is evidence to support these allegations, but once it is recognized that what are being sought by the authorities are some general presumptions, attention turns to the help which can be given by economic analysis. Unfortunately, economics is largely unable to provide the decision rules for policy which it is asked to give; although we hope to show that economists' analyses can question some of the currently fashionable dogma in this field.

The first problem for economic theory is that it developed an elegant analysis of equilibrium under conditions of perfect competition which was not strictly relevant to the real world. The predictions of this analysis could be used to show that private enterprise maximized the public interest, at a time when the aim of intervention was to convert noncompetitive into perfectly competitive situations. This purpose existed when it was thought possible, and when some limitations of the analysis, such as private versus social costs, were ignored.

In addition to perfect competition, economists developed other market structures, such as monopoly industries where all firms had some limited monopoly powers, and, lastly, oligopoly.

Oligopoly describes industries which are dominated by a small number of large firms. British examples are the car, detergent, and tobacco industries. Where the industry is small, the dominant firms will themselves be small. This may be important as economics attaches different characteristics to large, as opposed to dominant, firms. This distinction is not drawn in most public policy and will be largely ignored here. Another complication is that the economic theory rests on an implicit assumption that firms produce one product in only one market. Obviously, in practice, firms are never single product units; large firms may have diversified into a variety of industries.

Important points about oligopoly are, firstly, that it approximates to many real world situations, secondly, that economic analysis is unable to make the general predictions possible for other market forms and, thirdly, that this form of market structure can claim different advantages for the public benefit, *eg*, competition through innovation and quality of output rather than price. Public discussion usually lumps monopoly and oligopoly into the 'monopolized' industry and compares the resultant performance with that of a 'competitive' industry.

Before considering the light which economic analysis can throw on this comparison, we must note the importance attached in practice to the strength of the market structure and the means used to bring it about and maintain it. Some monopolies which exist technically are recognized as conferring little actual power. For example, the British Railways monopoly is restricted by road haulage, airlines, and the private car. Some monopoly industries may be protected by barriers to the entry of new firms; thus the NCB has, in practice, a legal monopoly of coal mining in the UK. In other cases the few firms in the industry may be well aware that any attempt to exploit their position would encourage new competition. Public policy, *eg*, the Monopolies Commission, has always attached great importance to any 'barriers to entry', whether they follow from patent protection, tariffs which restrict the effectiveness of imports, the financial and technical problems of entering at a large enough scale to be competitive etc.

Methods of obtaining and keeping monopoly power are also objected to; this is fundamentally on ethical grounds. Thus there was public disapproval of the British Match

Corporation's[1] bribing an expert called in to advise potential new entrants to the market. The bribe was effective; the expert advised against competing with the BMC. This is a particularly difficult area; economics cannot help in ethical judgements. Thus Courtaulds favoured its own subsidiaries rather than outside customers for artificial fibre. Is this unfair, or reasonable business practice? Most managers are bound to have met a situation of this type. 'All competition is unfair if you can't meet it' conceals the fact that like all other forms of human activity, business raises ethical problems.

Elementary economic analysis has demonstrated to innumerable students that monopoly, compared with perfect competition, will produce higher prices, a smaller output, and monopoly profits. This comparison, however, assumes that in both situations firms seek to maximize profits whereas it is open to any accused firm to claim that it limits its profits out of altruism – or fear of new competition. Secondly, the general predictions are only valid if costs are assumed to be the same in both cases – the monopolist may claim greater benefits from economies of scale. Lastly, the whole analysis ignores any relationship between market structure and technical change.

What evidence is there regarding the performance of monopolized, as opposed to competitive, industry? It should be remembered in what follows that the analysis is an amalgam of points concerning the effects of more or less competition, and of large firms against small firms. As has already been pointed out, there may be great competition in a monopolized industry while a small firm in another situation may have considerable monopoly power.

Price information is easily available, but difficult to interpret without data on structure and costs. On many occasions, an investigatory body, such as the Monopolies Commission, is forced to use international comparisons as the only way of assessing the effect of monopoly. Obviously, these may be confused by basic international differences in costs of production, or the Commission may have to compare monopoly with monopoly, but they provide valuable evidence for a particular industry. As a result, no general conclusion can be drawn; in fact, the only general statement is the economist's – that in the absence of competition, price will be higher than the marginal cost of production. This means that the money price distorts the true cost.

If a monopolist's price is higher than would be the case under perfect competition, the output would be restricted; since a higher price means fewer purchasers. Thus a decision to restrict output (and thus sell at a higher price) will be identical in its effects to a decision to charge that high price (and thus sell the smaller output). This is a very common strategy for firms; note that it is only available to those with some monopoly power. Whereas if competition is great the only decision is how much to produce at the market price, *eg*, the situation facing a farmer. There are many examples of restriction of output in order to obtain a higher price, *eg*, the marketing of diamonds.

The aim of monopolization, and the most common public objection to the result of this market structure, is the earning of high profits. As with prices, the information for the industry being considered is readily available, although there may be disagreement as to how profits should be measured. The problem of what is a 'reasonable' level of profit is, in the last analysis, normative. Firms have claimed that their 'high' profit rates should not be compared with the average results of other firms since part of their case is that it is their greater efficiency which enables them to be more profitable. Again, a Monopolies Commission may be driven to make international comparisons or to compare profitability between industries. High profit margins may bring public opprobrium in particular cases. Are there any general relationships?

Firstly, there are studies which show that there is a tendency for profit rates to increase with concentration,*

* Briefly, the proportion of an industry's output controlled by the largest firms. A 'high concentration industry' would thus have a high proportion of its output from a few firms.

although these results are very sensitive to the degree of potential, as well as actual, competition.

This is what we might expect. On the other hand, an investigation comparing profits, as a rate of return on capital and concentration, in British industry in the early 1950s showed that there was no tendency for high concentration industries to earn high profit ratios. There are, however, two points to be borne in mind here. Firstly, in so far as the valuation of a firm's financial capital will reflect the expected profit from those assets, we might expect that profits expressed as a percentage of capital employed might not vary as both profits and the valuation of the firm's resources have a tendency to move in the same direction. Secondly, there is evidence in this sort of study, as well as in others which compare profits with size of firm, that dominant and large firms have a better expectation of earning average rates of profit. Thus at a moment in time, large firms have a greater certainty of earning average profits, and over time their profits fluctuate less than those of small firms. This is also true when we compare profits earned by large and small firms in the same industry.

What evidence is there about the relationship between cost levels in monopolistic and competitive firms? This has a great bearing on profits and prices. Since profits may be expressed in relation to costs, the cost *level* is important. Two firms may both be receiving 20 per cent margins on costs but, for one, costs may be 100 (and therefore profits 20, price 120) but for the other, costs are 150 (and therefore profits 30, price 180). In so far as prices in practice are calculated as a percentage on costs, as in the above example, costs affect prices. Public policy has placed great importance on costs and efficiency, particularly when governments believe that by a Prices and Incomes Policy they can exercise some control over prices and profits.

At first sight, the advantages are likely to lie with large firms, since they are able to exploit economies of scale. There is abundant evidence to show that some manufacturing costs fall per unit produced as the size of production unit and length of the manufacturing run increase. Examples are steel blast furnaces, oil refineries etc. But it can also be shown quite easily that such economies of scale are not produced indefinitely and that there is a maximum efficient size of plant. Only very rarely does the manufacturing economies of scale advantage operate so as to allow only a very small number of firms in a domestic industry. Giant firms are typically multi-plant organizations; with each plant producing at costs no greater, but often no lower, than their single plant rivals.

There are, of course, other costs besides manufacturing ones and it is probably here that large firms obtain their greatest advantage. These would be the financial costs of borrowing money, marketing, and management. Large organizations benefit from marketing economies in their widest sense; a national chain of petrol stations presupposes a large oil company; advertising rates are lower to large users.

But this type of cost differs from manufacturing costs in two significant ways. Firstly, lower advertising rates are a benefit to the firm but not necessarily to the community – since the result may be merely a redistribution of profit between advertiser and advertising medium. Secondly, there is strong evidence that marketing costs are related to industrial structure in the sense that oligopolistic industries compete in this way rather than price. Table 19 shows an oligopolistic industry where selling costs are very important. So, whereas any real economies of scale in manufacturing will be in the public interest, this may not be so for lower advertising costs with increased size. Firstly, these cost advantages may be financial and not real; secondly, although the costs may fall, this change may be from a higher level, itself determined by the oligopolistic structure of the industry.

The relationship between structure and management is even more difficult to pursue, partly because management is not simply a cost but a prime determinant of what will be produced, how, and where; nor can we simply say that good managers produce good profits. Managers, then, are not merely a cost of production, they control costs and revenues. More particularly, they vary enormously. This can be held to favour the large firm which can afford to attract the best managers. This assumes that managers receive salaries; clearly there would be great attractions for a good manager, who could obtain the resources, in controlling his own business and earning profits as well as salary.

TABLE 19

Analysis of average Retail Prices of Detergents in percentages

	Unilever		Proctor and Gamble	
Factory Cost		46·0		43·5
Research, Administration, Distribution		7·5		7·0
Selling:				
Advertising	11·0		10·0	
Promotions	6·0		8·5	
Market Research	1·5		0·5	
Other	4·5	23·0	4·0	23·0
Total Manufacturer's Cost		76·5		73·5
Manufacturer's Profit		7·5		10·5
Manufacturer's Realized Price		84·0		84·0
Retailer's Margin		16·0		16·0
Retail Price		100·0		100·0

Source: Extracted from the Monopolies Commission Report on Supply of Household Detergents, 1965.

On the other hand, no resource of a business is more transient than its managers. They may die suddenly; they may move to other firms – and since the evidence is that most management moves are within an industry, they then become rivals. Furthermore, large firms are typically multi-product organizations and it is in large diverse organizations that the worst managerial problems arise.

Our last relationship between structure and performance concerns technical change. In the last thirty years this has become a widely debated matter, largely arising from writers[2] who have claimed that the realistic analysis is not one of equilibrium, beloved by economists, but of the process of growth. Thus any evidence regarding prices, costs, and profits is almost irrelevant. Growth involves change, more particularly the use of new technology and the development of new products. If it can be shown that large firms are more effective producers of technical change than their smaller competitors, or, if it can be proved that oligopolistic industries produce more rapid change, then this would be a powerful case for government intervention since the evidence is that most increases in output per head, and thus national growth, over the last fifty years have come from new technology rather than increased capital per worker.

Before considering the relationship between industrial structure and technical change it is necessary to define the latter term. In our definitions both invention (an increase in knowledge) and innovation (the commercial application of invention) are part of technical change. Research and development is one form of activity which may lead to technical change, but the latter is a wider term, since in our sense it includes new products and new processes in addition to increases in scientific knowledge which finally have a commercial application. Thus technical change may result from 'blinding flash' inventions, may involve virtually no costs to a firm (when it makes use of knowledge developed by others), and it may include new management techniques, *eg*, the mass production of cars associated with Henry Ford.

The original statements on this subject merely stated that large firms were the usual sources of technical change. This has been extended to cover both the large firm and industries dominated by large firms (*ie*, oligopoly). Table 19 shows how unimportant research seems to be in one oligopoly, but the most usual evidence produced to support the argument is the type in Table 20, which shows that large firms devote more resources to R & D than small ones. But this is not really enough if we are concerned with technical change in our wide sense – which is the form in which it provides a major element in growth.

TABLE 20
R & D and Firm Size

Size (in terms of employment)	USA (1958)			UK (1959)		
	5,000+	1,000–4,999	less than 1,000	2,000+	300–1,999	less than 300
Percentage of Enterprises doing Research	89	50	4	90	58	18
Percentage of Total R & D Expenditure	85	8	7	93	6	1

Source: *Science, Economic Growth and Government Policy*, OECD, 1963, page 87.
Quoted in *Concentration or Competition, a European Dilemma?*, McLachlan, D. L., and Swann, D., PEP, 1967.

Firstly, research shows that there is a weak positive relationship between R & D and size and that there is a threshold effect, *ie*, below a certain size firms undertake no R & D at all. But very large firms do not seem to devote proportionately more resources in this field than simply large ones and in some industries (*eg*, the American steel industry), the largest firm has a worse record than its fellow oligopolists.

Secondly, R & D expenditure as shown in Table 20 is a measure of an input; from the public interest standpoint we are interested in output or input/output comparisons. One immediate problem here is that some results (*eg*, new model styling changes in cars) may be held to be of little social value – although they add to costs and prices and may be profitable to firms in the industry. We will not pursue this point, which inescapably involves value judgement about the benefit to consumers of widened choice and pleasure from novelty. In practice, several studies in this field have used the (inaccurate) measurement of output of number of patents, when large firms score well. But if we distinguish between 'important' and 'unimportant' patents the evidence can be summarized in the phrase that 'large industrial laboratories are likely to be minor sources of major inventions'.[3] No study has ever contradicted this result.

We are concerned, however, not simply with structure and new invention but with its association with the speed with which inventions are taken up by industry. Here the evidence is that new technical knowledge diffuses faster the less concentrated is the industry. In the steel industry the basic oxygen furnace is agreed to be the only major advance in steelmaking this century. It was invented in 1950 by a small Austrian firm, the process was introduced into the USA in 1954 by a firm with 1 per cent of the market but it took a further ten years for the largest three firms in America to follow. In the meantime they are estimated to have built 40 million tons of obsolete capacity. We have taken this overseas example, partly because it is well established (*NB* it would appear that our reaction time, though better than the American steel industry, lagged behind others, such as Japan), also because the effects have been apparent. Whereas once the USA exported five times the amount of steel it imported, recently the ratio has been reversed. Here is one case where an industry has involved society in inefficient resource use by being slow to respond to new technology.

Evidence in this field is incomplete and was often not designed to test propositions which are important for policy makers. Persuasive arguments can be put forward to explain why large firms are in the forefront of technical change. They have a motive, the exploitation of change to yield monopoly profits, and the opportunity, with retained profits available for investment and sufficient size to be above any threshold. Since large firms are typically diversified, motive and opportunity refer to a wide field. On the other hand, it can be argued that large firms may be inclined to play safe. Two pieces of evidence support this view. Firstly, financial analysis shows how large firms seek to reduce fluctuations in their dividend distributions. Secondly, as we have already mentioned, their profits fluctuate less than those of smaller firms. Stable dividends and profits are not *public* policy objectives; their attainment is evidence of 'playing safe' rather than a fast pace of technical change.

There is much still to be learned about the factors which affect invention (*eg*, is invention a random occurrence or can it be predicted?), innovation (particularly its social value), and technical change. Many of the points presented above are still being debated, but most economists do not feel that the evidence for large firms, or dominant firms, as powerful engines of technical change is convincing. There is suspicion that views are often based on such reasoning as: the USA has the highest standard of living, this comes largely from technical change. The USA has most of the world's largest firms and has many highly concentrated industries. Therefore, if we want a higher standard of living we should imitate the structure. This rests on assumptions which we have already questioned. The social scientist would add the following comments: the USA was a great importer of technical change during her fastest growth, other factors are known to affect growth, other countries such as Japan are growing at faster rates without being large producers of technical knowledge. American industry is not significantly more highly concentrated than British industry – although American firms are of course larger.

What action are governments taking to alter or control the structure of industry? We will outline two administrative devices used, the Monopolies Commission and the IRC, consider their objectives, and finally comment on their results in the light of the foregoing analysis. Some parts of the policy have been ignored in what follows. Legislation concerning restrictive practices has clearly affected relationships between firms in an industry – and may have made mergers more popular. We have also ignored the recent beginnings of a policy actively to help small firms.

We have a Monopolies Commission, set up by the

Monopolies and Restrictive Practices Act 1948, and amended in 1953, 1956, and 1965. The Commission comprises eminent businessmen, trade unionists and academics, with powers to consider and report on an established monopoly in an industry, defined as where one third or more of the output of an industry is controlled, directly or indirectly, by one firm. Where a merger is contemplated so that the '⅓ Rule' would apply or where the assets taken over exceed £5 million, the Monopolies Commission (since the 1965 Monopolies and Mergers Act) may be asked to report on whether it recommends that the merger should be allowed to go ahead. The Commission has no powers of initiative, it only considers references made to it by the Board of Trade.* A reference usually asks the Commission to state whether or not the conditions of the legislation apply (*ie*, '⅓ Rule', £5 million assets Rule), whether conditions in the industry act against the public interest and, if so, then to suggest remedies. Its recommendations carry no legal weight; its report may be accepted or rejected by the Board of Trade which, in any case, may decide to follow the lines of action suggested, do something else, or do nothing at all. The powers available to the Board of Trade have expanded with the legislation and include: stopping practices which are considered undesirable; controlling prices; forbidding mergers, and breaking up merged groups.

Turning for the moment to the Commission's impact on existing monopolies of scale, it must be said that although major industries have been examined during the past twenty-one years, the total number of references has not been great. This follows from the small scale of the Commission and the fact that most of its reports are detailed, tightly reasoned examinations of the facts which take a long time to complete. For long periods it has appeared more radical than the Board of Trade, with the result that its full recommendations for action have rarely been implemented.

The crux of the whole procedure has been the problem of identifying 'public interest'. This concerns:[4]

* Currently, the DEP.

a. The production treatment and distribution by the most efficient and economical means of goods of such types and quantities, in such volume, and at such prices as will best meet the requirements of the home and overseas markets.
b. The organization of industry and trade in such a way that their efficiency is progressively increased and new enterprises encouraged.
c. The fullest use and best distribution of men, materials, and their industrial capacity in the United Kingdom.
d. The development of technical improvements and the expansion of existing markets and the opening of new markets.

These are obviously broad and non-specific guidelines. Under '*a*' the Commission is involved in detailed examination of profits, profit rates, and the components of cost. Where possible it makes inter-firm, inter-industry, and international comparisons. Evidence is also obtained of the availability of supply of the output. Technical change, export performance, and ease of entry into the industry are relevant under '*b*' and '*d*'. These are all qualities of an industrial structure* which have been discussed above and from which the conclusion can be drawn that there are no *general* economic predictions for, or against, market structure where one-third or more of the output is produced by one firm. This is reflected in the legislation, as the procedure is an investigation to obtain information which can be compared with the public interest criterion.

While the Commission can be accused of lack of imagination, slowness and reluctance to use new analytical techniques, in general it is considered to have produced consistent and thorough work. But its impact on industrial structure has been slight. To some extent this is the inevitable

* With the exception of export performance – which is presumably the result of the relative competitive efficiency of British industry compared with the rest of the world. Economists do not usually consider structure and export performance to be connected in any way that has not already been explained by economies of scale etc. The probable advantage of the large group is in having larger financial reserves.

result of an approach which means a single look at a small number of industries. The lack of support from the Board of Trade has also hindered its impact. In 1948 great importance was attached to the effect of publicity on firms' policies. We doubt whether there is much fear of a Monopolies Commission investigation. How many of our readers, we wonder, could outline the criticism of Courtaulds or indeed any other major report on an established monopoly in the last five years? So far as prices are concerned, popular attention has shifted to the work of the National Board for Prices and Incomes which seems capable of more rapid investigations and conclusions on a reference, and which can in certain situations keep prices under continuous review. (At the time of writing it is the intention that these two bodies should be merged.)

The Commission is also concerned with reporting as to whether or not a proposed merger* or one which has taken place within the last six months should be allowed to continue. It comes into operation in cases where a merger which falls within the scope of the 1965 Act fails to satisfy Board of Trade requirements. Since 1966 we have had an Industrial Reorganization Corporation with, as a major function, the promotion of mergers. There is no necessary inconsistency here since, as we have seen, we have no general criteria as to whether or not one market structure is preferable to another. Before continuing with a description of how this policy is carried out, we must consider in what ways attitudes to mergers might differ from those to established large firms.

A merger is obviously one form of growth for a firm, and official policy is presumably based on the premise that whereas it is generally impracticable to control internal growth which may lead to monopolization,† it is possible to anticipate such elements in external growth, *ie*, by merger. The '⅓ Rule' and '£5 million assets minimum size' are not

* Or takeover, *ie*, any situation where an independent business unit is absorbed, which fulfils the market share or assets rules outlined above.

† Except by MC references when the situation has already occurred.

intended to raise a presumption that any merger which qualifies is likely to be against the public interest.* What are the objectives laid down for those who carry out the policy? Public interest is not defined for the Monopolies Commission in the 1965 Act. It is the Board of Trade around which the whole merger policy revolves, since the Board does the preliminary sifting from 365 to 12 and maintains communications with other government departments and the IRC. It has developed its own criteria of 22 questions which depend on the type of merger.†

The IRC can foster mergers where this will promote industrial efficiency, profitability, and assist the economy of the United Kingdom. They look both for the advantages of large size which we have already discussed, plus extra objectives such as the quality of planning in the merger, the extent of any rationalization which will result, its effects on employment and on the Balance of Payments. We will examine each of these in turn.

Not only is a merger only one form of growth, but growth is only one objective of a firm – others are profitability, efficiency, and so on. Furthermore, in so far as economic growth is one objective of society, it may come without any given firm growing, but as a result of the birth of new firms in new, or old, industries. Like any other policy of a firm, it involves resource use and has an opportunity cost in terms of other actions which have to be sacrificed. Badly planned, poorly executed mergers are expensive for society. Evidence is starting to appear that these costs may be considerable in terms of short-term decisions, taken simply to ease a merger, and lack of managerial ability to obtain the potential benefits.

It is not uncommon for post-merger performance to fall short of the rosy forecasts made in the pre-wedding day battle. The IRC has always attached great importance to

* This is reflected in the figures that for the years 1965–8 over 3,000 large companies were taken over of which 365 fell within the 1965 Act. Of the latter, 12 were referred to the Monopolies Commission.

† And which are set out in an informative review of this policy; *Mergers: a guide to Board of Trade practice*, HMSO, 1969.

good management; this seems sensible if we look upon it as a necessary prerequisite for a successful merger rather than that good management is a conclusive motive for merger.* As we have pointed out before, managers are mobile and mortal and the evidence concerning the continued excellence of the 'management dynasties' is not encouraging.

Secondly, we have the importance attached to rationalization, *ie*, the closure of inefficient and excess capacity. The motive behind the merger is crucial here. Whereas a monopolist may well keep inefficient capacity operating by adjusting prices, so that even high cost units can make a profit, a merger in the public interest may well streamline an industry. In a sense that is all a merger can do, re-structure an industry, since although a firm grows by merger an industry does not. This is, of course, one of the advantages to the firm of growth by merger. As we have said, where an industry grows it can expect to find that prices will have to be reduced to clear the market. In the case of a merger, total supply may be unaffected.

In addition to mergers, the IRC is also concerned with rationalizing industrial structure by encouraging the transfer of operations between companies. Rationalization may well involve redundancies and these may create serious, and well publicized, unemployment situations in a locality. The IRC is bound to consider these regional implications – although we do not know to what extent. The Board of Trade, in consultation with other government departments, now obtains information from firms concerning the size and timing of redundancies and frequently requires assurances that estimates will not be exceeded and that properly negotiated procedures with trade unions will be followed, before deciding that a proposed merger need not be referred to the Commission.

* The authorities' view has been caricatured as 'In order to achieve industrial efficiency, find the most efficient firm in Britain and merge the rest of them into it'. See *Britain's Economic Prospects*, Caves, R. E. and Associates, G. Allen & Unwin Ltd, 1968, page 321.

The emphasis on the Balance of Payments effects of mergers is puzzling, apart from special situations, such as the social effects of American control of a large part of the UK car industry and the consequent 'adverse' remittances of profits; or the costs and benefits of international mergers where the motive is the exchange of information that may produce technical change. Admittedly, large firms may have advantages in being able to maintain marketing facilities overseas. But internal growth is another road to size and cooperative facilities are a possibility. Just how large is it necessary to be? Not all evidence unequivocally favours ever-increasing size nor is much known about the proportionate cost reductions that can be expected. The BMC/Leyland merger was advanced on the advantages to be gained from combined overseas marketing; but many other suggestions have been made to explain the relatively poor export performance of the UK car industry, *eg*, lack of attention paid to marketing at any scale of operations. No attempt seems to have been made to compare how great a favourable Balance of Payments effect (and whether this involves increased exports, reduced imports or whatever) is necessary to compensate for other less desirable results.

The IRC has operated since early 1967 in fostering mergers and reorganizations which it feels are desirable. It may do this merely by bringing the parties together for discussions or it may lend funds to make the merger viable. It has funds of £150 million of which £60 million have been used. It is prepared to favour one party in a contested bid, *eg*, its 25 per cent shareholding in George Kent Ltd; it takes a view about a preferred pattern of industrial structure, *eg*, it is at present manoeuvring, through its complete share ownership of Brown Bayley, to establish a strong special steels group in the private sector. Through the directorships which it now has on eight boards, and by 'follow up' interviews with completed mergers, it aims at intervention over a period of time. But in any case, many of its effects on the structure of British industry will be permanent, or very long-run, as few mergers are ever dissolved, although they may be eroded by new competition. But the more concentrated the industry, the less likely that competition will come from new entrants and the greater the importance of competition from other industries.

All this is a great responsibility, judged in terms of social costs and benefits rather than private sector profitability, and it is disturbing to find this part of the policy controlled by an organization which stressed in its first Annual Report that it was 'created by government but directed by businessmen' and recorded with pleasure that after two years of operations it was able to pay a dividend.

On the other hand, we have mergers being discouraged by a system which is basically controlled by the Board of Trade. The Board applies criteria which have not been the subject of legislation, before passing a small number (3–4 per cent) of cases to the Monopolies Commission. While some mergers (*eg*, Imperial Tobacco/Smiths Crisps) have been dropped at the preliminary stage, only four of the twelve before the Commission have had an adverse report.* It is too early to judge whether further action under the 1948 Act may be necessary to check whether the forecasts and assurances of behaviour made before a merger are being carried out.

To conclude this section of the chapter we must re-emphasize that intervention of the IRC type involves an assumption that the market mechanism will not bring about the desired results – or perhaps not in time. Obviously other policies can be invoked to facilitate change, such as those which will be discussed in the next section. 'Tinkering' is an inaccurate word to describe the present policies if only because it implies ineffectiveness. IRC policy has had results that have been considerable. On occasions, such as in the trawling industry, the Monopolies Commission and IRC have taken different views as to what structure is desirable.

* Ross/Associated Fisheries, United Drapery Stores/Montague Burton, Rank/De La Rue, Barclays/Lloyds, and Martins Banks. Assurances about future behaviour have been given to the Board of Trade in other cases, *eg*, Chrysler/Rootes and GEC/AEI, sometimes at the preliminary stage and sometimes as the result of the Commission's findings.

The following quotation represents a widely held American view about the size of firms.

> 'In the last 20 years enough empirical data have become available to dispel the naïve belief, so fashionable in our age of innocence, that firms are big because they are efficient; that firms are big because they are progressive; that firms are big because they are good; that firms are big because consumers have made them big.'[5]

Together with the arguments presented earlier in this section, it makes the basis for much of recent British policy in this field appear rather thin and unconvincing.

More hopeful signs are a recognition by the Board of Trade that, in general, industry requires a pattern of size distribution of firms. Foresight is an uncommon gift among men and planning techniques have not advanced to the extent to which we can predict the industrial size pattern we will require in the future. Still less do we know which firms will be important. After all, IBM was a small firm when it developed computers, Xerography was turned down by large firms. Polaroid grew from an idea which was turned down by Eastman Kodak.

Investigations have resulted in our now knowing quite a lot about the implications of a pattern of size distribution of firms in an industry; but these are largely 'still photographs', rather than a cine film of industrial structure.

2 *Policies Concerning Resource Mobility*

The extent to which the resources of an economy can move, or be moved, is a most important factor affecting how well a country reacts to change. Impediments to mobility mean that resources may be misused because they cannot be transferred to situations where they can be used more effectively; this problem is only made more pressing by technical change. Mobility is of particular importance for the economist as he is trained to think in terms of optimal solutions to problems and is then confronted in reality with barriers to their attainment which often come from lack of mobility. To the manager, trained in an imperfect world, it may not be apparent that any extra mobility is possible.

By mobility, we mean not merely geographical movement, but a change of job or a change of industry. The distinction between the last two forms is that a resource may do the same job but in a different industry – and thus contribute to a different form of output. Mobility applies to all three factors of production. We will ignore land and concentrate on labour, after saying a little about the mobility of capital.

Government action concerning capital mobility refers to *financial* capital. Since the financial markets are better developed here than in other countries there is less need for some types of intervention – except in exceptional circumstance, such as the Capital Issues Committee which sought to ration long-term finance in the post-war period. There is continuing intervention to favour certain industries which face problems in attracting financial resources in the free market. The Shipbuilding Industry Board and the Agricultural Mortgage Corporation are examples. The Industrial Reorganization Corporation may be considered another attempt to divert resources in a particular direction. Finally, there are institutions, such as the Industrial and Commercial Finance Corporation and the Finance Corporation for Industry, which have been developed, with government support, to remedy shortcomings in the capital market, such as the problems of smaller firms in the first case and large sums of medium-term finance in the second.

For the rest of this section of the chapter we will confine our attention to methods of improving labour mobility. Policies *a*, *b*, and *c* are concerned with short-term techniques largely affecting the geographical mobility of a worker. Policies *d* and *e* affect mobility in a more general way and also seek to improve the quality of the nation's labour resource.

a. The Public Employment Service

This is the oldest policy concerning mobility. Labour Exchanges were set up in 1911. Designed to speed up reemployment and reduce unemployment by keeping a register of vacancies and those available for work, its development has suffered by its popular association with 'the labour' as the place where unemployment pay is arranged. A recent development has been improved information regarding vacancies in other areas as opposed to the largely local nature of the service – although we lack intensive services in areas affected by redundancies. The previous emphasis on manual work has been extended by a special service for office personnel. A start was made, in 1966, in providing an occupational guidance service for adults – something which has been carried out by the Youth Employment Service for some time. All these techniques are designed to improve the imperfect labour market.

In a recent article,[6] Dr A. P. Thirlwall suggests that benefits exceed costs for both the job-finding and the redeployment aspects of the services. For example, since the average cost of a placement is £6, and the average daily production is £5, the cost has been covered if the service reduces the period of unemployment by 1·2 days compared to what would have happened in the absence of Labour Exchanges. The likelihood is that the actual period by which unemployment has been shortened is greatly in excess of this.

b. Legislation to improve the status of the worker

This started with the Contracts of Employment Act 1964, which laid down statutory minimum periods for notice of dismissal, ranging from one week (26 weeks' service) to four weeks (5 years' service). More important for our present interest has been the Redundancy Payments Act 1965. This provides for the legal right of a worker, where the job for which he was employed has disappeared, to receive a tax-free, lump-sum payment, irrespective of any other benefits which he may receive. The amount depends on age, length of service, and average weekly earnings. Payments are made by the employer, who is reimbursed from a fund financed by increased employers' National Insurance contributions. Due to the high level of recent redundancies, the fund has been in deficit and recently the reimbursement has been reduced to a flat 50 per cent rate. Between December 1965 and August 1969 a total of £176·6 million has been paid in over 800,000 cases. At present the average award is about £250.

The result of this legislation, together with earnings-related unemployment benefit introduced in 1966 (so that a married man with two children may expect to receive about 60 per cent of his average recent working income), has been to make the redeployment of labour much easier than it would otherwise have been. Firstly, the majority of employers have felt that change involving redundancy brings less hardship to those concerned; this may be contrasted to those who dismissed all their employees and then re-engaged them in 1965 (this delayed the impact of the Redundancy Payments Act for which there is a two year minimum service qualifying period). Trade unions have been much more ready to accept redundancies, for in addition to negotiating about phasing of the run-down process they now know that, irrespective of any special redundancy compensation which may be agreed, their members are protected by a statutory right to benefit.

Apart from its social content, the importance of this legislation in reducing barriers to change has been considerable, although we know of no way in which this can be measured. But there is evidence, mentioned in Chapter Two, that the unemployed are remaining longer on the register. This may be due to the nature of recent redundancies, *eg*, more older workers becoming unemployed, but it is believed that an important consideration is that the unemployed are now in a position to discriminate more between jobs offered to them.

c. Assistance to geographical mobility

The three schemes at present operating have their origins in intervention designed to solve the regional problem in the 1930s. They are:

i) The Resettlement Transfer Scheme. This applies anywhere in the country to the unemployed who have poor employment prospects at home but who obtain a job elsewhere for which no local labour is available.

ii) The Key Worker Scheme. This is designed to assist employed workers transferred to key jobs in new or enlarged establishments in the development and intermediate areas.

iii) The Nucleus Labour Force Scheme. This concerns unemployed workers in development and intermediate areas who have been given jobs which involve temporary transfer to their employers' parent factory for training.

Under all three schemes, benefits include fares (to the job and up to six visits home per year), a lodging allowance when living away from the family (of £4 4*s* 0*d* per week) and, under the RTS scheme, once a family home has been found, assistance with removal expenses and the legal costs of house purchase. Similar schemes operate in other European countries and all were originally seen as ways in which, by improving geographical mobility, regional differences in employment could be reduced by migration.

It is now recognized that, by themselves, such policies cannot solve regional problems and attention has turned to the other types of action mentioned in Chapter Two. Partly this is because, even in quite desperate situations, workers are reluctant to sever local social ties and persist in preferring policies which bring work to them – and thus of course reduce the effectiveness of policies promoting migration. It is now recognized that migration has a chain effect. For example, the tendency is for the most mobile section of the working population to be the younger, better-trained workers (who may move not in response to unemployment but to reduced career prospects); this reduces the growth prospects for those who are left. The loss of a region's purchasing power caused by migration affects other earnings, *eg*, in the service sector. Finally, there is evidence that the present forms of assistance do nothing to counteract two most important restrictions to mobility. Assistance for house purchase is useless where houses are unavailable or very expensive – this situation is only recently starting to improve. Except in special circumstances, return fares to interviews are not paid; this seems a pity because ignorance of opportunities is still a great problem.

The assistance has been available for many years, yet outlays for all three schemes are in the region of £0·5 million *pa*. About 6,000 individuals received travelling assistance, the most used facility, in 1967. Various explanations have been put forward to account for the small impact of these schemes, such as lack of publicity and restrictive qualificatory conditions; but the fact remains that we have fewer resources devoted to this type of intervention here than abroad. This is important since, even if the case for inter-regional migration as a cure for regional problems is not accepted (and it continues to be forcefully advocated), geographical mobility is often a fundamental prerequisite for job or industrial mobility.

We now turn to consider policies on training, which are mainly concerned with job and industrial mobility. In the short-term much training is aimed at increasing the productivity of a worker at a given job and this objective may be shared by the employer and employee. But even the most narrow vocational training, by allowing a worker to progress further, expands his earning horizon; most training does more than this, it allows the individual, and if not the firm then society, to progress. A better trained labour force is therefore more easily able to react to change. Education has the same effect; it opens opportunities to the young which are wider than unskilled work or entry into the parent's trade.

d. The Industrial Training Act 1964

The objectives of this Act are i) to ensure an adequate supply of properly trained men and women at all levels in industry, ii) to secure an improvement in the quality and efficiency of industrial training, iii) to share the cost of training more evenly between firms. It was a reaction to criticisms of the British labour force on the grounds that many labour shortages were largely of trained men, that apprenticeship* which had formed the basis for Britain's relatively well-trained work force during the last hundred years was time-expensive and rigid, that modern training methods were essential to retrain an adult worker, that in certain industries, and for women generally, training was non-existent and that some firms suffered by operating excellent training schemes from which the output was attracted away by others whose strategy was high wages rather than the provision of training facilities.

How does it come about that differences arise between the needs of society and of firms in this matter? The basis for a policy does not rest on any economic assumption that improvements in the quality of the labour force are always to be welcomed. Of course there may be social reasons for wishing to do this – but these are usually stronger in the case of education than training. An improvement in skills is worse than useless if it produces results which are inappropriate for society. Not only have resources been misapplied, but those trained who cannot use their skills will feel frustrated. From the viewpoint of the economist, both the firm and the State can look on training as a form of investment. We will look at each decision in turn, which will demonstrate why there is a case for government intervention in this field.

For the manager, considering training as a form of investment has three advantages. Firstly, it focuses attention on labour as a form of capital (*ie*, a resource as in Chapter Two); one which may be improved as well as by self-improvement via experience, and one which may in practice deteriorate. Secondly, the identification of costs and returns allows a more rational decision to be made. Thirdly, the use of an investment appraisal approach makes it clear that the equimarginal rule, mentioned in Chapter One, applies here – so that the returns from the last £1,000 spent on training should equal those from any other investment.

The firm's costs will include: the provision of the physical

* Forty-two per cent of boys who left school in 1966 were apprenticed.

facilities for training; paying for the instruction, including lost production from an internal instructor; lost output from those being trained; lowered morale among workers not selected for training, and increased attractiveness of trained workers for 'poaching'. Benefits comprise: increased value of output from the trained workers (which should be discounted to a present value to enable comparisons with present costs, such as lost output); any production undertaken as part of the training process (which may be associated with lowered wages paid to trainees); increased morale; greater skills specific to the firm's techniques of production (which may make alternative employment less attractive); improved quality of recruitment, and any government grants or loans.

The calculation for society differs from that applicable to the firm in the following ways. Firstly, any 'poaching' is not a social cost, apart from any costs of transferring from one job to another. This emphasizes the reasons why firms are likely to be interested in more narrow and specific types of training. Secondly, where those being trained would otherwise be unemployed the opportunity costs of training will be zero. Thirdly, as a form of public sector investment, returns from training must be compared with those from other possible projects. Fourthly, training may help to achieve other, wider macro-economic objectives. Lastly, government policy which increases the volume of training may generate external economies of scale which will be shared by the private and public sectors. Training Boards change the firms' calculation by varying both costs and returns.

To return to the policy. The Act set up a Central Training Council and provided for the creation of Industrial Training Boards – of which there are, at present, 26 covering 15 million employees with an ultimate objective of 30 and 18 million employees covered. The CTC is responsible to the Department of Employment and Productivity for the overall operation of the legislation. It also has committees which deal with common problems – such as the Committee on the Training and Development of Managers. It sponsors research into training problems and produces reports.

The impact of the Act in industry has come from the operations of the ITBs. Each has, at present, autonomy on how it runs its training affairs but all have the power, after ministerial approval, to compel firms in their particular industry to pay a training levy. This is then reimbursed to firms in the form of grants which depend on the training undertaken.

ITBs have wide latitude in the size of their levy and the method by which it is collected. The first decision that has to be taken is whether to aim at a sum which will be able to reimburse all training undertaken by the industry or simply to have sufficient funds to provide an incentive for training. Secondly, they must decide whether to start on a limited scale or aim at collecting their target amount in the first year, even though they may initially be unable to distribute their funds as it takes time for firms to hear of training opportunities and to react to them. Most ITBs collect by a payroll percentage levy, but a few have chosen a *per capita* sum (*eg*, iron and steel, £18 *pa*). There is a recent tendency for differential systems depending on occupation or skill to be considered, especially where one Board covers a range of different industries, *eg*, ceramics, glass, and mineral products.

No consensus has appeared on the best method of operating a levy, nor is this likely since industries differ widely and several approaches to training are permissible. The Engineering Industry Training Board is the largest (with a levy yielding over £80 million pa) and is an example of a Board which started with a high levy (2·5 per cent of annual payroll).

Whereas the levy determines the resources available to a Board, the grant system largely determines the quantity and quality of training undertaken. Grant schemes vary so widely that few generalizations are possible except the following: they may repay training costs; may pay fixed amounts for particular types of training, and may, or may not, link the grant to a percentage of the levy paid by a firm. Several ITBs combine various schemes; the most complicated, as might be expected, is that operated by the EITB.

It is difficult to see why for a particular firm there should be any link between grant and levy, except as a check where there is doubt about the basis of either of them. Clearly grants must act as an incentive; the Industrial Training Act does not force firms to train their labour force and it is still possible to consider, as a minority still do, that the levy is simply a tax on labour.

But, in general, the Act has been welcomed by industry. After the initial fun of playing the new training managers' game of 'how high a percentage of levy we recover' (the record is believed to be over 400 per cent and there are legends of firms who have found training to be their most profitable activity); even our limited experience shows evidence of more training being carried out. On the whole, this has tended to be better training, although there is a lot of communication still necessary between industry and colleges carrying out off-the-job training.

There is still little evidence regarding the cost-benefit results of this legislation as a whole, although there are some studies for individual companies. One of these showed that 75 per cent of the net benefit was in the form of reduced labour turnover and the rest came from higher output performance. Certainly the scope of the policy is considerable – over £130 million was paid out in grants for the year ending March 31st, 1969. The very size of the operation has led to criticisms from industry of the considerable bureaucracy which has been created. However, administration accounted for less than £4 million of the £130 million.

Apart from the initial mistakes, inevitable in such a new undertaking, so that some Boards are increasing their levies, others reducing them, while patterns of grants change frequently, four major general problems have still to be overcome.

Firstly, few resources are directed to the training of women and girls. The arguments why this is so are familiar ones but, we would suggest, do not always reflect the current time pattern of work for women, *eg*, a tendency to marry at an earlier age and to resume work once family responsibilities allow.

Secondly, the reluctance of employers to give day release from work for training continues to lag behind the educational arguments in favour of this form – ITBs are believed to be having some success here, but all find this a problem. On the other hand the EITB in particular is having unexpected success with the equivalent of apprentice instruction operated on a modular basis.

Thirdly, there is the problem that small firms typically operate on a poor levy/grant ratio. This may reflect backward attitudes to training (*eg*, the rumpus in the Agricultural, Horticultural, and Forestry ITB), but also results from special problems, such as the difficulty of sparing trainees off the job, the lack of suitable courses, and geographical dispersion. Group training schemes are being developed to help here. In many cases, however, the small firm's problem is that it would like to return to the old system which allowed it to employ those trained by others.

Finally, there is the danger that if ITBs continue to develop at different levels and on different lines for too long, with centralized information about best practices, the Act could lead to industrial differences in training which may be as harmful in the long run as the inter-firm differences which it was designed to stop. This is most likely when the ITB has a narrow industrial base and where the emphasis is on specialized training – this could hinder mobility.

Our views on this legislation are bound to be coloured by our involvement in its results. The authors' first-hand evidence is heavily biased, since inevitably we tend to meet training managers who favour more, and better, education and training. We admit that there is little evidence, one way or the other, regarding the costs and benefits of the policy. But a brief résumé of the goals which could be attained shows the importance of this recent development.

More widespread and better training can help i) to reduce

unemployment, ii) by reducing geographically localized redundancies, help the regional problem – and thus perhaps reduce inflationary pressure, iii) finally, by increasing productivity, lead to economic growth.

e. Government Training Centres

Unlike the situation in France, it has never been the policy here that the government should be a major direct provider of training. GTCs were originally set up in 1917 to train disabled ex-servicemen, in the inter-war period they trained the unemployed, and trained nearly 0·5 million engineering and ammunitions workers in the Second World War. In 1947 there were 80 centres concerned, again, with rehabilitation. Whereas continued involvement on that scale may have been unnecessary, there have been considerable social costs involved in allowing the number to fall to 13 GTCs (capable of training 2,000 workers *pa*) by 1962 only to step up the policy to the present level of 42 centres (9,500 *pa*). The recent shortage of places has led to schemes for assistance (of both finance and staff) to firms who provide semi-skilled training on their own premises.

GTCs offer training in over 40 trades, with an average course length of about six months. Most training is in upgrading from semi-skilled to skilled grades. This initially led to problems of trade union acceptance, particularly when the trainee finishes with the equivalent to an apprenticeship level of craftsmanship. Various agreements have been made to overcome these difficulties, recognizing that trainees are as skilled as apprentices but may initially be slower.

Most applications for places have been from individuals wishing to retrain or upgrade their skills and this has been a considerable individual investment since grants for trainees are low. Prior to 1968, if a firm sponsored a trainee it had to pay his costs (unless, perhaps, it was in a Development Area). Since then, firms may, with local union approval, pay the wages of sponsored trainees (which they may be able to reclaim from their ITB) and not be charged the costs. GTCs have always been free to individual entrants.

Official policy has consistently given a minor role for GTCs – even now they provide places for less than 2 per cent of the unemployed. In the last three years emphasis has been placed on linking training programmes with local industry, siting new GTCs in problem regions and making special facilities available where local redundancies occur. But only a minute proportion of those made redundant seek GTC places (*eg*, 330 from 12,000 in a BMC scheme in 1966), although better chances to retrain are often mentioned by the unemployed as a shortcoming of the present system.

A recent survey of the costs and benefits of adult re-training shows a rate of return of 30 per cent. This high figure is influenced by two considerations which do not apply to private sector training. Firstly, the opportunity costs are low – and will be nil in the case of a trainee who would otherwise be unemployed. Secondly, the costs of running the scheme are offset, not only by benefits of increased mobility and efficiency, but in financial terms by reductions in unemployment pay and increased future taxation.

The five policies at present operating to improve labour mobility in this country are a mixed bag. Some have not been in operation very long. Their objectives vary and we are short of evidence as to how far they manage to achieve them. Taken together, they now account for about 1½ per cent of GNP, which is twice the proportion of 1960. A recent OECD report believes that some policies should have been introduced earlier, but we now seem to have techniques and a level of intervention which matches our European neighbours – and goes beyond them with the Industrial Training Act. In changing the economic environment in this way, we have created opportunities for individuals and problems for firms and trade unions.

3 Promoting and Shaping Technical Change

The last section of this chapter is concerned with government policy in shaping and promoting technical change. This is, of course, one of the objectives of intervention by the

Monopolies Commission and the IRC which we considered in the first section. What follows is intended to be read in conjunction with the earlier material; the same definitions, *eg*, of technical change, will be used. This section is less concerned with describing the institutions involved in this aspect of the manager's environment; our concern is more with reasons for the policy and the various forms of policy which seem appropriate.

Why do governments intervene in this field? Apart from a straightforward value judgement that it is desirable to promote pure science, there are two major reasons. Firstly, governments feel that they must keep up with the great technical changes in warfare – for which 'defence' is the current euphemism. For reasons which are a mixture of strategy and national pride, several large nations have policies of heavy commitment in this form. The economic implications of government defence spending will be mentioned later.

Secondly, technical change is recognized as being an important cause of growth – and governments are judged in the light of this criterion. That 'economic growth' and 'better living standards' are not necessarily closely related was explained in Chapter One. There are external costs of technical change, of which pollution is currently being emphasized, and there are the costs of change itself.

Much discussion in this field seems to assume that growth, change, and technical change are all the same thing. There are many examples of industries which have declined as a result of technical changes in which they could not share, horse-drawn transport and the railways, for instance. It is even possible to ascribe national decline, in the past at least, to new technology. However important, technical change is merely one part of the total changes in an economy. Some, such as the discovery of new mineral deposits or population increases, are only loosely connected with new technology; others, such as social changes, themselves generate invention and technical change. Thus in the 19th century it was public disapproval of women and children working in coal mines which brought about the introduction of equipment to replace them.

Social factors shape and limit technical advance. To be a Luddite or to 'stand in the way of progress' are terms used to describe minorities who wish to halt or divert some forms of technical advancement; it is important to recognize that society as a whole frequently does this. In many parts of the world there are maximum speed-limits on roads of about 60–80 mph. It is technically possible to produce family saloons to travel at twice the speed, but society does not wish it. Instead, public pressure in the USA has very quickly made various safety devices commercially practical which were already technically possible. When it is recognized that some technical advance is stimulated by the need to offset the undesirable effects of other technical change, *eg*, noisy jet engines require silencing, and that technical change is a flow concept – at any moment of time there is a stock of knowledge which has not yet been diffused – it becomes obvious that the popular view is too simple.

We have concentrated in this section on the relationship between technical change and growth because this is the one which is currently fashionable. In the wider view technical change is probably more important, not for the increases in GNP which it may bring, but for changes in the quality of life. There are today fewer dangerous, dirty, and servile jobs – men have been replaced by machines or new ways have been found of obtaining the output required. The possibilities for a population which has warmth, light, and shelter available when required, which is immensely more geographically mobile, and which has old demands satisfied, even if newer demands are created – all these are something more than can be measured in terms of increased GNP even at costs of stress, noise, and pollution.

To the extent to which technology replaces men by machines it has raised the fear of unemployment. At least fifteen years ago in this country gloomy predictions were made of this type – apart from Luddism. Since then, we have had a considerable amount of automation, but little permanent unemployment which has resulted from it. In the USA, where the process has been going on longer, *it has been estimated that to produce 1958 output with 1947 technology would have required a $\frac{1}{3}$ increase in labour input;* a similar result is likely over the last 12 years. But again the predictions have not been fulfilled; economic systems can accommodate this kind of change like any other.

The impression is often given that we live in times of unparalleled invention, research and development effort, and technical change. Let us examine the evidence in a general way; although clearly much depends upon whether we are comparing this decade with an earlier one, or 'the middle of the 20th century' with 'the 19th century'. Certainly, in this country and the USA, there has been an increase in the number of patents filed. All sorts of social factors have influenced this. We suggest that the quality of invention is very relevant; if the increase comes, for example, from household gadgetry the results would not be important.

It is easy to show that firms devote increasingly large resources to research and development, and that the proportion of the GNP of advanced countries allocated in this way has increased in the relatively short time span for which we have information. This is at first sight persuasive evidence, but requires qualification. Firstly, R & D expenditure is a measure of input and not output; we saw in the first section of this chapter how the resources used by large firms produce fewer major inventions than from their smaller rivals. A significant proportion of most national figures is devoted to military research. Compared with other forms of research, this leads to lower rewards in terms of commercial technical change, as opposed to strategic security. There is some 'spin-off' from space research – non-stick saucepans – but most observers find it disappointing, and our feeling (though we do not know of a study which provides evidence in either direction) is that it would compare badly with the results of commercial R & D. This is not meant to be a criticism of space or military research, which must be judged by other criteria.

Another problem in evaluating R & D input is that the available figures invariably lump the two together. Usually development is much more expensive than research but it includes innovation in the form of changing the style of the product, *eg*, the car industry. Our point is not that such changes are a form of competition which society may feel are rather unimportant, but that this expenditure is related to the structure of the industry. Industrial structure and competitive strategies change with time. Taxation which discriminates in favour of capital gains as against profits may make R & D more worth while from the point of view of shareholders. This would happen if the results of the research programme increased the size and future profit expectations of the firm, thus allowing the benefits to be taken in the form of higher share prices – and thus capital appreciation. Alternatively, it has been said that research has become a more socially acceptable activity than the earning of profits.

Historically, to associate research and production together in the same institution is quite new. Even 70 years ago firms, while they might be founded on a new technical development, once established relied much more on bought-in technology. The frequency of the examples of stress between successful researchers and the organizations for which they work show how the marriage may not be ideal. This can also be a problem at the development stage. Some large American companies have recently set up 'venture divisions' financed by a partnership of the firm, the innovator and a finance company, to promote riskier and more difficult development and innovation.

A widespread belief is that we are going through a period of rapid technical change. While the time-lag between invention and the new product has shortened, it is very easy to overestimate the extent of current change. Probably it is impossible to measure objectively the relative importance of the jet engine and the computer as compared to the steam engine and gas lighting.

More disturbingly, for those who advocate policies of

accelerated change through a 'white-hot technological revolution', or who believe that British research and development input must be raised 'in order to maintain international competitiveness', there has been no correlation between a country's growth rate and the proportion of GNP devoted to R & D. Partly this is because of the poor correlation between R & D and technical change – the latter being the ultimate objective for society. As has already been mentioned, basic scientific discoveries are quickly made available to others – they are an example of (nearly) free goods. So the USA imported European science and technology during the 19th century and Japan has done so since 1947. It is new products and new processes which can be quickly applied to existing markets which tend to be closely guarded secrets. Even such path-breaking new technology as the transistor became quickly available to a number of firms.

Finally, R & D effort does not simply lead to growth (or vice versa) because investment is connected to both of them. In Chapter Two we looked at the complex investment/growth relationship; clearly there are also costs between R & D as a type of investment – it may be in the form of time spent in looking at a problem, or getting together a research team, rather than the, often heavy, costs of buildings and equipment. But, like investment, it requires the use of resources which would otherwise be consumption goods.

The results of the programme may call for heavy new investment (*eg*, steel above). Secondly, in so far as most measures of technical change for a nation include the advantages from 'learning by doing', a part of the total change will be the result of past investment levels.

Having tried to explain the complexities of the economic aspects of technical change, we will turn to a list of the policies which have been tried in this country. Finally we will consider suggested causes, and solutions, to British problems in this field.

Up to the 19th century a widely used technique for encouraging invention was by offering cash prizes, medals etc. The ship's chronometer was invented in this way. The tradition is carried on by the Queen's Award to Industry which has been awarded for technical innovation. A much older incentive, and one still used universally, is the granting of patent rights. The economic effects of any given patent system is a well written up field of economic theory; interest has always been particularly strong in the USA where patents provide one of the few legally permissible forms of monopoly. The public interest has, therefore, to take two opposite forces in account.

Firstly, there is the incentive effect of the monopoly gains which may be obtained; there are instances of individuals who have decided that to invent is the best way of raising their living standards (*eg*, Xerography), other inventors are inspired by non-economic forces (*eg*, Wankel). Although the rewards from a patent can be distinguished from other forms of monopoly on the grounds that they (exclusively) reflect work which has been done, society may suffer from restrictions on the diffusion of new knowledge resulting from a patent. As a result there have been various suggestions for changing the system; these range from forcing patent holders to grant licences at reasonable royalties to all applicants, to rewarding inventors by means of a lump sum rather than by any subsequent benefit from the use of the invention. The National Research Development Corporation (see below) was set up with helping the private inventor as one of its objectives. But criticisms still remain, such as the complexities of taking out a patent – the Patent Office is being reorganized – and the treatment of royalties as income for tax purposes, rather than as a capital gain, as in the USA.

The fundamental problem still remains that of reconciling patents as a spur to inventive activity and as a barrier against new competition. As W. Kingston[7] has pointed out, irrespective of the motives of the inventor, monopoly power, which brings the ability to earn above normal profits, is necessary for an invention to be exploited. When Tolstoy repudiated the copyright on his work no one would publish him.

Two policies aimed at promoting R & D and technical change have already been dealt with. The questionable basis for promoting industrial concentration *to attain this objective* has already been examined. The only clear way in which such a policy would increase R & D inputs would be from merging firms so that they grow beyond the minimum size for which R & D is possible (the 'threshold effect'). For example, it has been calculated that R & D costs are typically only 10 per cent of the total costs of putting a successful innovation on the market. On the assumptions that firms do not usually spend more than 5 per cent of their sales on R & D and that a small research effort costs £100,000 *pa*, then we have the result that firms must be prepared to spend £1 million (£100,000 × 10) on an innovation and must have sales of £2 million *pa* (£100,000 × 20).

Secondly, the system of investment grants now in operation provides a considerable incentive both to increasing R & D effort, by subsidizing the investment required and by encouraging investment generally, and increasing the rate at which technical knowledge is embodied in new productive capacity.

Over the last twenty-five years, most government policies in this field have involved the direct allocation of public funds to R & D, assistance to existing institutions, and the creation of new institutions designed to overcome particular problems. We will deal with each policy in turn.

The most important single use of funds has been in defence contracts – principally in the aircraft and aerospace industries. This form of public expenditure, financing work carried out in the private sector, is held to have special advantages since any new technology developed can be used by the companies involved for their commercial operations. The USA is held up as an example of a country where this technique has been used successfully – subject to the qualification concerning 'spin-off' already mentioned. In the UK the allocation of contracts has been used as a method of enforcing structural change, *eg*, the consolidation of the aircraft construction industries into two large groups carried out in the early 1960s. Recently the technique has been extended to purely commercial development, *eg*, the placing of contracts worth £6 million for the construction of advanced machine tools which are subsequently lent for evaluation by users for periods of up to two years. This has not been particularly successful.

Grants for research at universities have been another policy, the extent of which has recently been criticized. Much of this argument has been between those who feel that academic research should be more closely linked to the economic and social needs of society, and their opponents who hold that the pursuit of knowledge is an end in itself, which has nothing to do with these objectives and cannot be the subject of cost-benefit analysis. In our view universities are, in practice, concerned with both sets of objectives; undoubtedly some research is 'useless', and many universities have poor links with industry.

We now turn to the institutional framework which has been set up by the government. An attempt has been made to rationalize control over these institutions by the setting up of the Ministry of Technology in 1964, emphasized by the Science and Technology Act 1965. As a result, 'Mintech' is now responsible for:

a. Direct government assistance to industry of the type mentioned above.

b. Science Research Councils; concerned with university research.

c. Research Stations; of which the oldest is the National Physical Laboratory. Some have been criticized as being too small or insufficiently alive to the commercial implications of their work. As a result, some small stations have been closed down and Mintech has encouraged others undertaking work commissioned by private industry on a fee basis. Most research stations are primarily concerned with defence work. In this field they have a special role as institutions which can carry out basic research objectively in the national interest. This is obviously the case where secret work is concerned; it also applies to research on social problems, *eg*, noise.

d. Research Associations; there are over 40 of them undertaking cooperative research for member firms in their industry. They obtain finance from their members and from public funds. Started in 1918 and greatly developed after the Second World War, they were originally seen to have great value as a way of helping small firms past the R & D 'threshold'. However, they have grown only slightly compared with the enormous increase in R & D expenditures by firms; as a result they suffer from the problem of small size.
e. Mintech is now responsible for a variety of types of intervention, some of which (such as the IRC and the operation of the IDC scheme) have been dealt with elsewhere. One very successful institution which now comes under its control is the United Kingdom Atomic Energy Authority. The decision to have a single government agency responsible for the civil uses of atomic energy, while leaving the contracting work in the private sector, is held to have resulted in the present international advantages which we have in this field.

Another part of Mintech's operations is the National Research Development Council; since this is directly concerned with the process of development and technical changes we will consider its work at greater length.

The NRDC began operations in 1949 and now has £50 million borrowing powers and invests at about £4 million *pa.* It

> 'promotes the adoption by industry of new products and processes invented in government laboratories, universities, and elsewhere, advancing money where necessary to bring them to a commercially viable stage. NRDC speeds up technological advance by investing money with industrial firms for the development of their own inventions and projects'.[8]

There is an overall requirement that its work should be in the public interest. As a result it is not interested in gadgetry or 'gimmick' patents.

The NRDC receives proposals from three sources: private individuals, firms, and government agencies. After examination of a proposal it may become involved in one of two ways; either seeking licences for a patent or providing finance for the development of an idea.

Apart from the products of government or university research, for which it is the usual channel whereby they are patented, the NRDC has received over 8,000 proposals from private inventors during the last 20 years. Of these, less than 100 have been supported; a low proportion, but in keeping with experience elsewhere. In all, it has about 500 licensing arrangements for patents which are earning revenue. This part of its operations may be summarized as: sifting proposals, which range from the important to the crackpot; advising inventors to take patent protection, and using its very wide industrial contacts to find firms who are prepared to make use of the patented invention under licence. Conscious of its non-commercial objectives, the NRDC prefers limited, but non-exclusive, licensing arrangements, *ie*, it prefers the advantages of potentially wider use of the invention to higher earnings that could come from a patent monopoly. It also favours British firms.

The second aspect of the NRDC's work consists of supporting 'development projects'. These are fewer in number, but may require very heavy financial commitments, *eg*, over £5 million for hovercraft, and include the more revolutionary technical developments, *eg*, dracones, fuel cells, enzyme production. In 20 years it has supported about 450 development projects, of which 120 have been unsuccessful, 80 are earning revenue (of which one has produced over £1 million in royalties), and 250 are still being developed.

There are two ways in which it is prepared to assist. Firstly, it may provide finance for development work in return for the major share in any royalties received. Secondly, it may enter into a joint venture with a firm in which a loan or an equity interest is provided.

How effective has the NRDC been? It very quickly demonstrated that there was no source of private invention, dammed by the restrictive policies of large firms, which could be tapped to release a flood of new ideas. By its nature it has tended to have received proposals which are too risky or with too long a time horizon for the private sector. This is, of course, one of the NRDC's important attributes; as has already been mentioned the public interest may be involved in a time profile of benefits which is too long to be commercially worth while. Not only may it be prepared to wait longer for a return, but it tends to be less interested in managing a venture than would be the case with a commercial 'venture capital' organization.

Two of its limitations, a shortage of finance and an interest in development only to the stage at which firms are prepared to take over (*eg*, the setting up of the British Hovercraft Corporation), are inherent in its constitution. It has also recently attracted criticism on the grounds that by insisting that it should hold the patents of any accepted proposal and by claiming a large share of any royalty payments, it drives too hard a bargain with the private inventor. On the other hand, it could be argued that, as the NRDC provides a safety net for development work which would not be sponsored by the private sector, it must earn good returns on successful projects to subsidize its failures. Its success rate, given the special circumstances already mentioned, compares well with similar organizations in other countries.

Finally, we will turn to some suggested explanations of the alleged, poor British performance in the field of technical change. Alleged, because some criticisms are based on reasoning such as 'we have a poor growth performance, R & D is important to growth, therefore our R & D performance must be poor', which we have shown earlier to be an over-simplification. Furthermore, there is no consensus of opinion about explanations.

One explanation blames British managers for being poor developers and users of basic research. We will take two examples which have been used to support the case. Various surveys have shown examples of British inventions which have been commercially exploited elsewhere; in particular there is evidence that the USA has exploited successfully many results of basic research from other countries. Secondly, we have a favourable 'technological balance of payments', defined as the net result of royalties paid for the use of overseas investments compared to those received for British inventions. This 'distinction' we share with the USA, in contrast to other European countries and Japan. This may seem a point in favour of the success of our technology; however we would suggest that it can be more usefully considered as evidence that British industry is less prepared to make use of new technology developed elsewhere. While the effects on the 'real' Balance of Payments are small, those on industrial development may, as in the case of Japan, be very large.

Why, to use the jargon of a couple of years ago, do we have a 'Technological Gap' between ourselves and the USA? A common explanation has been that the USA has a higher proportion of engineers and scientists working in industry. More recent evidence shows that, in fact, Britain and European countries have a higher proportion – the difference between management lies in the higher educational standards of the average American manager. So the explanation may be found in the quality of labour resources looked at earlier.

In so far as it is all education, rather than just scientific education, which is important, then the changing demand for education away from pure science in favour of social science and arts subjects, current among secondary school children, may be less important than has been predicted. In any case, it would be quite legitimate to consider it as an example of a social restriction on technical change. The adult population may think that it needs more scientists for a 'white-hot technological revolution', the younger population may have different objectives.

A recent study[9] persuasively argues that British technological development suffers from two other problems. It argues that government intervention should redistribute our R & D resources in favour of high technology industries while at the same time reducing the proportion of resources devoted to the aircraft industry – which it shows offers a poor return in the face of American competition. It also claims that our R & D labour resources contain two few engineers and too many scientists.

In conclusion: managers face a changing environment, one cause of these changes is the result of the application of new technology. The importance of this is advertised by large firms, heavily committed to R & D expenditure, and by governments hoping to promote faster growth. At present we know far too little about all the relationships involved to have an agreed policy about what should be done. We are still clearing the motes and beams from our own eyes. Clearsightedness will not solve the policy problems, however, since the conflicts of objectives will remain. Oligopolistic firms will continue to see innovation as the important competitive strategy, governments will seek to deliver growth, inventors and researchers will remain misunderstood and in conflict with the institutions which provide them with resources. The public will remain a mixture of Luddites, conservationists, and revolutionaries.

References

1 'Report on the Supply and Export of Matches', Monopolies Commission 1953, para 145.
2 Schumpeter, J. A., *Capitalism, Socialism and Democracy*, Allen & Unwin, 4th Edition 1961. Galbraith, J. K., *The New Industrial State*, Hamilton, 1967.
3 Hamberg, D., 'R & D Essays on the Economics of Research and Development', Random House, 1966, page 69.
4 Monopolies and Restrictive Practices (Inquiry and Control) Act 1948, Section 14.
5 Peltzman, S., and Weston, J. F., eds., *Public Policy Towards Mergers*, Goodyear, 1969, Chapter 2 by W. Adams, page 26.

6 Thirlwall, A. P., 'On the Costs and Benefits of Manpower Policies', *Employment and Productivity Gazette*, November 1969.
7 Kingston, W., 'Invention and Monopoly', *Woolwich Economic Papers*, No. 32.
8 NRDC, *Inventions for Industry*, No 32, July 1968.
9 Peck, M. J., in Caves, *op. cit.*, Chapter X.

Bibliography

Caves, R. E., and Associates, *Britain's Economic Prospects*, Brookings Institution, 1969, covers this chapter as follows:
First Section – Caves, Chapter VII.
Second Section – Caves, Chapter VIII.
Third Section – Caves, Chapter X.

Hunter, A., ed., *Monopoly and Competition*, Penguin Modern Economics Series, 1969.

Mergers: a Guide to Board of Trade Practice, HMSO, 1969.

Institutions such as the NRDC, the ICFC, and the FCI will supply on request booklets describing their operations.

Jewkes, J., Sawyers, D., and Stillerman, R., *The Sources of Invention*, Macmillan, 2nd Edition 1969.

Parsons, S. A. J., *The Framework of Technical Innovation*, Macmillan, 1968.

CHAPTER SIX

The Private and Public Sectors

In Chapters Four and Five we considered some of the techniques available to the government for intervening in the economic environment. In this chapter we are interested in assessing the relative sizes of the private and public sectors of the economy. In practice we will be concerned with the nature and extent of the public sector, since an analysis of the structure of industry is outside the scope of this book. Our measurement of size for the public sector will be to compare it to GNP. After a discussion of the principles on which government activity is based, and some of the objections to the techniques used, the chapter sets out some measurements of the size of the public sector. This is followed by a survey of the factors which determine its size, followed by some international comparisons. Finally, attention focuses on the characteristics of the UK tax system. The Appendix sets out in some detail an economic analysis of the Selective Employment Tax.

It is obvious that, at a given moment of time, an individual's attitude towards the form and size of the public sector will be largely the result of his political and social value judgements. Attitudes to government intervention (and thus to the public sector) may be polarized between the *laissez faire* view that the economic role of government is to hold the ring and let business get on with it, and the socialist view favouring public ownership of the means of production and exchange. Even in these extreme cases 0/100 or 100/0 relationships are not advanced theoretically or found in practice. Thus State provision of armed forces was part of any *laissez faire* theory and the Public Sector/GNP proportion, by one measurement, was higher in 1800 than 1900. On the other hand, some non-State enterprise exists in communist countries and a considerable private sector is part of socialist Yugoslavia.

At present, contention in the UK does not concern either pole, but the intermediate positions. What is now at issue is a comparatively small expansion or contraction; this sort of problem is susceptible to the economist's marginal approach, when the advantages and disadvantages of a small change are examined, although the criteria used in reaching a conclusion will be political and social as well as economic. And this makes logical discussion very difficult, since what is in question is not merely the size of the intervention but also the techniques used. Now, as we pointed out in Chapter One, techniques (*eg*, monetary policy) are ways of carrying out policies and these policies should logically be consistent with ends. The ends, however, are political and it can be said safely will not be agreed on, still less ranked consistently, by the pundits at a club bar. We will try to describe and measure the public sector; the reader's own social and political prejudices will colour the spectacles through which, like us, he looks at the picture.

One attitude to government intervention is that it reduces freedom, either by dictating certain courses of action (*eg*, drive on the left in the UK), or reduces choice (high taxation stops holidays in Bermuda). In the case of driving on the left, we assume that motorists are so alive to the benefits that no one would disagree – this is a regulation which is obeyed and accepted, in contrast to speed limits which are sometimes ignored and often not accepted. Apart from a debate about which side of the road we should all drive on, the only discussion likely about 'driving on the left' would be about the way in which the regulation was promulgated – a Minister of Transport who caused 'Keep to the left' signs to be erected at 100-yd intervals along every road would be accused of wasting money. Now most economic intervention involves the possibility of discussion on both these grounds; the policy and its form.

To a reader who feels that any loss of his freedom of action is insufferable, we can only point to the driving type of example – and point out that the taxation which curtails his holidays may provide a subsistence living standard for an old-age pensioner. Clearly this is in the OAP's interests, although we have no scientific way of saying whether this redistribution increases the welfare of the economy as a whole.

This lack of scientific precision has not stopped society being generally agreed that in some cases, such as the unemployed, or old-age pensioners, *some* redistribution should take place. This may be in the form of transfers of income, or entitlement to increased consumption, *eg*, welfare services. A great deal of almost random redistribution has resulted from inflation. The economist's position is that *he* cannot demonstrate that £1 taken from a millionaire and presented to an old-age pensioner is a transfer which increases the welfare of the economy. Where the interference with individuals' course of action may restrict the choice of one group but widen that of another the problem becomes completely political – comprehensive education is an example of this.

Another objection may be that the chosen form of intervention has treated an individual harshly. Thus, if a new road has just reduced the value of a householder's main asset he is likely to feel unsympathetic to road-building programmes. Or a farmer who feels that the agricultural price review does not favour his pattern of output may believe that some other system of support to farmers would be preferable. Governments obtain revenue and spend in order to change the economy and any change will favour some rather than others. Unfortunately, since most regulations are general in their nature, it is likely that some will be unfairly treated compared to others as a result of imperfect administration or execution. These injustices appear in the private as well as public sectors; we naturally require higher standards in these matters from the latter.

Another personal, subjective view hostile to intervention

is that in this way economic power is removed from the impersonal market mechanism and concentrated in the hands of policy makers and executives. While allegations of corruption and partiality are not often made against the Civil Service in the UK, most people, perhaps subconsciously, agree with Lord Acton's phrase 'power corrupts and absolute power corrupts absolutely'. Although the executive authority of the Civil Service is enormous, its traditions result in policy stemming from Parliament while its size and diversity mean that to view it as a centralized organ of power is unrealistic. There are many concentrations of (economic) power in the private sector which, although they are smaller in size, are much more closely controlled – this forms part of the popular case for controlling monopoly.

The impersonality of a freely operating market is considered very valuable by some – often those who complain of the 'faceless civil servants of Whitehall'. If we look to effects, however, then for example, a given price increase may be the result of monopoly power or a higher purchase tax. Both reduce the range of choice open to the customer. On the other hand there is a difference in the mechanism which gives the result. The consumer may feel that if a (monopolist) firm puts up the price he can buy elsewhere or not buy at all. The second alternative is open still if the increased price is caused by the imposition of a tax but the decision to levy the tax is the result of the voting process – which means that usually about half of the electorate disagree with the political opinions of the Chancellor's party[1]. Furthermore, the collection of a tax is accompanied by legal sanctions.

Having pointed to some of the reasons why a given amount of government intervention will represent different things to different people, we now turn to the problems of measuring the size of the public sector in the UK. The major problem here is that such terms as 'size of the public sector' and 'public expenditure' can be defined in a variety of ways. This is not mere pedantry; the manager should recognize that the different definitions can produce results which, expressed as

TABLE 21

A) '*Public Expenditure*' (*eg*, as in *Blue Book*, Tables 49 and 50). Defined, in its broadest sense, to include current and capital spending of central and local government, together with capital expenditure and debt interest of nationalized industries (N1). See Table 22.

B) A LESS Government receipts of rent, interest, and dividends.
Surpluses of public corporations.
Central and local government borrowing from the private sector (N2).
= B '*Public Expenditure which has to be financed from taxes, rates, and National Insurance contributions*'.

C) A LESS National Debt interest payments (N3) (*eg*, interest on War Loan and Treasury Bills).
Social Security payments (N3) (*eg*, payments for sickness or unemployment and old-age pensions).
Grants and Subsidies (N3) (*eg*, student grants and investment grants)
= C '*Public Expenditure on Goods and Services*' (N4).

D) C LESS government investment (N5)
= D '*Government Consumption*'. See Table 23 for international comparisons.

E) A LESS National Debt interest (N6).
Capital expenditure of nationalized industries (N7).
= E '*Public Expenditure as used in forecasts and planning documents*'.

F) A LESS Capital expenditure of nationalized industries.
Local government expenditure.
National Insurance Funds (N8).
PLUS Grants and lending inside the government sector (N9).
= F *Supply Estimates*. This is the expenditure figure controlled by Parliament in the annual Vote on Account (N10).

N1 Note that, although this measurement gives the largest money total, it includes the nationalized industries only in so far as they are investors and pay interest on earlier borrowing. The size of a public company would never be measured in this way. Either of the usual accounting measures of size, turnover, or capital employed would give a still larger total. Furthermore, a nationalized industry which did not invest would not enter into this measurement at all. Debt interest is also highly arbitrary in that legislation wiping out debt (which has been done) would affect the result.

N2 An increase in government borrowing will, of course, reduce the size of B while increasing the National Debt and interest payments to be made in future. The nature and extent of borrowing is a technique of intervention and is also affected by institutional factors.

N3 These are all examples of 'transfer payments', *ie*, the re-allocation of purchasing power between economic units, such as the use of taxation, to reduce the taxpayer's spending power and increase that of the unemployed.

N4 This measurement represents the government's claim on real resources (materials and factors of production). A change in C alters the quantity of goods and services available for private consumption or investment. The inclusion of transfer payments in A and B obscures the real effects of intervention when A or B are expressed in relation to GNP, since, in the latter measurement, transfers of all types (*eg*, private transfers from wage earner to an indigent aunt) are ignored to avoid double counting income.

N5 We have already pointed out how arbitrary the decision to classify an item as capital formation or expenditure may be.

N6 Excluded here since it is largely the result of past borrowing.

N7 Excluded since 1965 as being determined by the commercial requirements of the particular industry in relation to the whole economy.

N8 Treated for control purposes as if they were a commercially operated insurance businesss although in practice in all countries (Table 23) such funds require assistance from general taxation.

N9 Largely Rate Support Grants from central to local government. To include such a transaction in a measure of public expenditure would be to double count both the income to the local authority and the expenditure of central government. This is an example of a transfer payment which is always conventionally excluded.

N10 Clearly this is a measure of little economic significance, which may change in the opposite direction from A–D (*eg*, by a shift in the balance between local authority rate income and expenditure). It is included here because, being debated at length in Parliament, it receives annual publicity. Recognition of the shortcomings of the present system of presenting public expenditure and the inadequacy of the Parliamentary Vote on Account procedure has led to official proposals for change. These are outlined in a 'Green Paper' (*Public Expenditure: A new Presentation*, Cmd 4017, HMSO 1969). The objective is to present current figures and projections for measurement E (and thus to bring back the capital expenditure of nationalized industries into the debate. See N7).

N11 The analysis would be of net items so that, *eg*, national insurance contributions will be deductible from national insurance payments. Further analysis would distinguish between claims on resources, transfers, and capital items (but differing from our treatment to exclude payments which caused an alteration in the size of available resources. Thus student's grants, since they affect the employed population would be classified under resources). Estimates for future years would be part of the presentation but would distinguish between figures that were merely projections and others which were the result of policy decisions.

a percentage of Gross National Product, vary between 20 per cent to over 50 per cent. An attempt at reconciliation of some of the more popular measures is given above in Table 21. An analysis of A is given in Table 22 and international comparisons of D in Table 23 later in this chapter.

All the measures given are in terms of expenditure, although B uses a definition of expenditure which reconciles with the major sources of central and local authority's revenue. Clearly to add together income and expenditure when determining the size of the public sector would be to double count. Yet in discussions, increased taxation and, say, increased social security benefits are linked together as being two examples of increases in the public sector.

Table 21 is an attempt to simplify the explanation of some rather tortuous reasoning by giving the outline of a variety of measures, A–F. Some of the principles, as well as the details, on which a measurement is based are set out in the Notes (N1–N10).

Certain conceptual problems should be borne in mind when considering these figures. Firstly the 'public sector' is heterogeneous; the distinction between transfer payments and command over real resources is particularly important. Thus a decision to cut expenditure on education might be carried out by reducing student grants, employing fewer teachers, or building fewer schools. The first change has little effect on resource use,* the second and third involve fewer real resources being used by the public sector on consumption and investment respectively. Or to argue in reverse sequence, a decision to hold total expenditure constant but to shift the component parts by reducing defence spending and raising pensions achieves a 'balance' only in a limited accounting sense. In real terms the government is reducing its claims on resources.

Some changes in expenditure may be the result of a posi-

* On the assumption that this resulted in a marginal change in the number of students which was insufficient to free real educational resources for use elsewhere. There would, of course, be a long-term effect on resource use in the economy resulting from a less well-educated labour force and a short-term effect of increasing the labour supply.

TABLE 22

An Analysis of Public Expenditure in the UK, 1968

		£ million	£ million
1	*Collective Services Provided by the State*		
	Defence and External Relations	2,776	
	Education	2,176	
	National Health Services (including local welfare, child care etc)	1,999	
	Social Facilities (including water, sewerage, refuse disposal, libraries, fire service, employment exchanges, roads and public lighting, transport and communication)	2,709	
	Law and Order	424	
	Finance and Tax Collection	223	
			10,307
2	*National Insurance & Pensions*		3,344
3	*Direct Government Intervention*		
	Housing	1,115	
	Other Industry and Trade (including Investment by Public Corporations)	1,863	
	Agriculture	400	
	Research	178	
			3,556
4	*Debt Interest*		
	Central Government	1,249	
	Local Authorities	543	
	Public Corporations	123	
			1,915
5	TOTAL PUBLIC EXPENDITURE		19,122
6	Gross National Product		36,686
	% Total Public Expenditure (5/6 × 100)		52·1%

Source: *Blue Book*, National Income and Expenditure 1969, Tables 1, 49, and 50.

tive decision, such as investment by nationalized industries. Others result from changes which are uncontrollable given a decision about the level of services to be provided, such as an ageing population increasing the amount paid in pensions.

Distortion comes from the basis of measurement. We have

already seen that GNP can be measured at market price or factor cost. In both cases the basis for the private sector is the revenue earned by firms, either before or after indirect taxes. But much of the government sector, for example the Health Service, is valued at cost and not price. The market price valuation of the output of the NHS is incalculable but certainly greater than its cost. While we are not suggesting that the NHS ought to earn a profit, we are pointing out that the valuation at cost is on a different basis from that which would be used in the private sector. We measure the size of the detergent industry by its revenues, not its costs – the difference being profits.

Another difference in the public sector will be the relative absence of advertising, as a cost to be recouped in revenue. Of course, the nationalized industries advertise and may do so in competition with each other. This expense would not be recorded in our measure of the size of the public sector. Traditionally, however, little advertising is done by the social services – which is not to deny that the effectiveness of rate rebate or supplementary benefit schemes may not be improved by better public understanding as a result of informative advertising. The economic effects of advertising are widely disputed, by economists among others. It is agreed, however, that in some market situations advertising will cause prices charged for the output of an industry to be higher than would otherwise be the case and thus cause the private sector to be that much larger.

Whether we use cost or market price, however, money is used as the measuring rod. It is obvious that in times of inflation we cannot compare one year's figures with another in real terms without adjusting for changes in the value of the measuring rod. But inflation also involves some distortion between sectors, since prices charged for public sector output are likely to increase less rapidly than the average. This may be the result of greater public awareness and resentment of such price changes; *eg*, public lavatories have charged 1*d* for many years and increased postal charges provoked an outcry. Partly it results from the fact that price restraint is administratively easier to enforce in the public sector.

Lastly, the size of intervention, by any measure, is greatly affected by its form. Thus the replacement in 1966 of investment allowances (which involved a reduction in tax paid) by investment grants (involving a transfer payment out of taxation to firms who qualified) caused a considerable increase in public expenditures. Partly this was the result of an actual increase in the size of the incentive but to a considerable extent it simply reflected the change in technique. Under the new system there was an increase in expenditure, not merely a reduction in income.

In all measurements except A, the size of the public sector diminishes if a service once provided free is then run commercially. Prescription charges reduce the size of the public sector.

The layman, struggling to grasp the implications of a measure which seems applicable for his purpose, may feel by now that he has had enough. His confusion is often based on the belief that when comparing the public and private sectors, or the public sector to GNP, he is dividing a national cake. This is not so. The cake varies in size according to the measurement used. Every statistical measure of GNP involves assumptions about what should be classed as production, *eg*, the work of housewives is ignored. The treatment of transfer payments (Table 21, N3, and N4) affects the totals. More fundamentally, as we saw in Chapter Three, one objective of government intervention is to influence the level of economic activity and also growth, that is to say they help to determine the size of the cake.

It should now be clear to the manager that there is no one measurement of the size of the public sector; it is therefore difficult to express accurately the problems and relationships between the private and public sectors.

Before making international comparisons of the nature and extent of the UK public sector, we will have a look at some of the explanations which have been advanced to explain changes (largely increases) in this sector over time.

We warn the reader that there is no 'model' describing public/private sector relationships. Some hypotheses exist which can be used to make general forecasts. What we do have, however, are insights into what has happened which are more fruitful than simple claims that a particular political party seeks to increase while another tries to reduce the scope of government intervention.

Our first approach to the problem is via welfare analysis, developed in the 1920s and 1930s and associated with Professor Pigou. This considers the basis for a transfer of economic activity from the private to public sector using the concept of community welfare. To take the simple case of a new service provided by the government out of increased taxation, this would increase the community welfare if the marginal increase in utility from the service obtained by the taxpayer exceeded the disutility of the increase in his taxation. This analysis therefore offered situations in which a rational taxpayer would welcome an increase in the government sector, *eg*, where he paid more tax but received a still larger benefit. On the other hand, it could never offer decision rules to policy makers as there is no way of assessing the changes in personal welfare which result. Furthermore, it is a purely economic explanation which cannot take account of changes in the opinions of society. This is an important limitation, since it would seem that a major factor in the extension of such types of intervention as the social services and State ownership has been public acceptance.

A second approach is that of 'Wagner's Law',[2] developed at the end of the 19th century and revived recently in this country. Basically the 'law' states that, notwithstanding any attempts at financial control, public expenditure in advanced economies will increase at a faster rate than national output. Wagner divided State intervention into three types. Firstly, the maintenance of law, order, and the commercial framework in which business operates. This, he thought, would increase at a faster rate than output because of the increasing complexity of the economic environment. Latter-day examples of this type of interference would be regional policy and the work of the IRC. Secondly, he pointed to the State provision of services, which he held would grow as only a State corporation would be large enough to withstand trade cycles and make use of very large scale production processes. The third category includes the provision of services, such as communications and education. Growth in this sphere would result from the prevention of private monopolies and recognition that State provision was necessary where benefits would not be measured accurately by the market mechanism. At the time when Wagner formulated his 'law', Western Europe provided many examples and his hypothesis was designed to explain this phenomenon. The UK and other advanced countries still show the same result. The limitation of the theory is that it seeks to explain only growth, and not the level of the public sector, and that it considers only the long term.

The third approach to the determination of the size and nature of the public sector results from Peacock and Wiseman's work, the only major recent investigation into the question for the UK. Their results show a general increase in government expenditure during the period; but not a steady increase – there is a strong tendency for peaks to be reached at times of social upheaval (particularly the two World Wars) followed by a plateau as the expenditure falls – but not back to the previous level.

The general trend can be attributed partly to the 'concentration process'. As economies develop, the service sectors tend to become more important in relation to the primary (agriculture and mining) and manufacturing sectors. Thus improvements in communications will tend to concentrate intervention in central rather than local government. This may follow from greater public knowledge generating a demand for the better services which they come to know exist in other parts of the country, *eg*, holidays away from home made people aware of the better environments away from the old urban centres. It is also because services are forms of activity in which the central government, by direct provision or through nationalization, has an advantage in relation to private enterprise. It is not simply chance that most nationalized industries can be classified as a service or public utility. New techniques make not only the large-scale provision of services possible but also allow large centralized administrative units to perform efficiently.

The 'peak and plateau' pattern of growth is explained by Peacock and Wiseman using a 'displacement effect'. They believe that in times of settled social conditions government expenditure will tend to grow, if at all, only so far as tax receipts increase with the growth in GNP per head. Greater increases in government receipts will not be possible because of widely held views about what is a 'tolerable' tax burden. No scientific definition of 'tolerable' can be given and it remains a matter of opinion.

The exact nature of the relationship between growth in terms of GNP or GNP per head and taxation obviously depends greatly on the tax structure. Three examples will be given here which the reader can relate to the description of the tax structure below. Firstly, where direct taxes are progressive,* there will be a tendency for tax revenue to increase more rapidly than GNP per head. An example would be the yield from a continuously progressive PAYE income tax. Another steeply progressive tax is Estate Duty; here the increased yield from greater wealth has to be offset by more successful techniques of tax avoidance.

Purchase taxes may be expected to increase at the same rate as rising expenditure on the commodity. Whether this happens depends partly on whether the tax is 'specific', *eg*, so much per lb of tobacco, or *ad valorem*, *eg*, purchase-tax rates on the wholesale price of an item. In the first case the yield relates to the physical *quantity* sold, in the second to its *price*. Under either method the increase in tax paid may be faster or slower than the growth in GNP depending upon the nature of the item taxed. Thus, increased living standards leading to a rapid increase in car ownership will cause a greater than proportional increase in receipts from Road Fund licences. On the other hand, spending on tobacco has

* See below for a definition. No value judgement is implied.

not risen recently in the UK so that the excise tax yield from this source has grown slowly even though the tax rate has been increased. Thirdly, so far as the yield from income tax is concerned, it is money, and not real, increases in GNP that matter. Since, in this country, the income categories in which either no tax, or less than the standard rate, are paid, are changed only rarely, inflation means that more and more income earners now pay the standard rate.

To return to the idea of a 'tolerable' tax burden – which both for an individual and for society as a whole may be quite different, and probably lower, than the desired level of expenditure* but which limits government intervention – it seems that this may be greatly changed in wartime. Modern wars necessitate considerable increases in the public sector. Munitions and the pay of the Armed Forces require funds financed out of current taxation and borrowing – this provides the largest single source of increases in the National Debt (see Chapter Three.) So, as a percentage of GNP, expenditure on the National Debt was 11 per cent in 1900, 5·6 per cent in 1918, 4 per cent in 1938, and 4·4 per cent in 1950. Furthermore, the social upheaval has, in the past, brought to public notice the bad living conditions among the poorer sections of the population. Thus evacuees in the Second World War demonstrated how many of the community were badly fed and poorly housed. When the war is over, the public sector typically falls to a plateau. The public, having accepted higher taxation and new forms of taxes as being inevitable and in the national interest during the emergency, has had its notion of a tolerable tax burden shifted to a higher level. At the same time, the desirability of increased expenditure on health, free milk for schoolchildren etc, is accepted and becomes very difficult to stop.

This analysis is highly persuasive even if the size of the effects cannot be calculated precisely. It does not claim to

* Thus a taxpayer may be in favour of a more expensive Health Service but feel that he cannot bear an increased tax burden. Society may favour increased defence expenditure but feel that the government sector measured by income should be reduced.

be comprehensive; and our account of Peacock and Wiseman's arguments makes no such claim.

Two further areas for investigation are how it can explain peacetime periods of rapid growth in public expenditure* and the extent to which it relies on an assumption that governments, including politicians and civil servants, have an objective to increase rather than decrease the size of the public sector. This contrasts with the 18th-century and 19th-century view of government that its objective should be to hold down expenditure – sometimes expressed as the Gladstonian objective of public parsimony.

We now turn to an examination of some of the distinguishing features of the public sector in the UK – some of which have already been described. The international comparisons in Table 23 show that the UK ranks sixth out of the eight advanced nations for which we have figures, using this particular measure of the size of the public sector. Columns 2 and 3 measure expenditure on real goods and services whereas cols. 4–7 concern transfer payments. Civil expenditure (col. 3) is very much in line with most other countries although significantly less important than in Sweden and greater than in France and Switzerland. Defence expenditure (col. 2) is shown to be proportionately higher than elsewhere – with the exception of the USA.

Turning to transfer payments, subsidies (col. 4) and foreign aid (col. 7) are generally unimportant.†

Interest on the National Debt (col. 6) is considerably higher in the UK than any other country. Bearing in mind, firstly our comments on the 'burden of debt' in Chapter Three, and secondly that we are about to combine a real item with a transfer payment, if we add cols. 2 and 6 we find a reason for the present size of UK government expenditure. No

* Just before 1914 public expenditure was rising at a rate which, projected to 1955, would have been very similar to the actual position without the displacement effect of two major wars.

† It is of interest to note that, although advanced nations belonging to the United Nations have pledged themselves to give foreign aid up to 1 per cent of GNP, only France has reached the target. The performance of Scandinavian countries is poor.

other country, except the USA, approaches this proportion of GNP spent on defence and debt interest, both of which reflect the international responsibilities undertaken by the UK now and in the past.

TABLE 23

International Comparisons of Government Current Expenditure as percentages of GNP at Market Prices

Country (ranked by column 8)	Consumption of goods and Services		Sub-sidies	Social Security benefits and other transfers to persons	Interest on National Debt	Foreign Aid	Total Current Expendi-ture
	Defence	Civil					
(1)	(2)	(3)	(4)	(5)	(6)	(7)	(8)
France	4·1	9·4	2·2	17·6	1·2	1·0	35·5
Netherlands	3·7	12·5	0·7	15·5	2·8	0·2	35·4
Sweden	4·5	16·4	1·7	10·9	1·5	0·3	35·3
Norway	3·5	13·4	4·5	11·1	1·2	0·2	33·9
West Germany	3·9	12·8	0·8	14·7	0·9	0·5	33·6
UK	5·8	12·1	2·0	8·5	4·0	0·5	32·9
USA	9·3	11·4	0·2	5·8	1·3	0·3	28·3
Switzerland	2·5	9·3	1·0	7·2	1·3	0·2	21·5

NB All Figures are for 1967, except Sweden and Switzerland 1966.
Since GNP is measured at market prices the percentages cannot be reconciled (but are generally lower) than those of Table 24 when GNP at factor cost is used.
Since the measure is current expenditure, investment is ignored.
Source: OECD's National Accounts Statistics. Adapted from Table III in Hill, T. P., 'Too Much Consumption', *National Westminster Bank Quarterly Review*, February 1969.

Finally, we turn to spending on social security benefits (col. 5) where great differences appear. While the UK spends more in this way than the USA or Switzerland, we lag far behind Continental countries. Whatever may have been the situation in 1948, the social security benefits received by a UK citizen, such as family allowances, unemployment pay, and sickness benefit etc, are now very low by the standards of our two major European competitors, France and Germany. To some extent real differences are concealed for this item and for Civil Consumption (col. 3) by institutional factors. Thus in France there is no National Health Service in our sense. Payments made by individuals are reimbursed from Social Security so that the French percentage (in col. 5) should be reduced by this amount (about

4 per cent) to be comparable. On the other hand, NHS expenditure, about 3·5 per cent, is included in Civil Expenditure (col. 3) for the UK and the removal of this item would widen the gap still further between the amount spent here and on the Continent on administration, education etc.

We argued earlier in this chapter that no single 'correct' measure of the size of the public sector exists and that anyway the individual's assessment of a statistic will be subjective and depend greatly on the form of the intervention. Nevertheless, the brief analysis of one measurement above should, at least, dispel an illusion that our public sector is high in relation to the countries with which we normally compare ourselves, *eg*, USA, Germany, and France. This is true also if we take any of the components of the public sector with the exception of defence and interest on the National Debt. There is a popular belief that we have an army of civil servants; the evidence is that they would be no more, or less than, a fair match for those of other advanced Western countries.

We now turn to taxation – the aspect of government intervention which at all times and in all countries has been the most contentious aspect of the public sector. We have already looked at the effect of taxes on prices but we do not have the space here for treating two other important aspects of taxation, namely what is a good tax in terms of its social effects (*eg*, should taxes be based on the ability to pay?) and what are the full economic implications of the various forms of taxation. The Appendix to this chapter looks at the Selective Employment Tax in some detail. This is for two reasons. Firstly, we felt that an economist's analysis of *one* tax may be of interest. Secondly, no recent new tax has provoked quite such an outcry, based on an unusually high level of ignorance at best and at worst ill-concealed special pleading.

Our general remarks will be centred on the international comparisons given in Table 24. Economists' analysis usually distinguishes between two main types of taxation. Firstly, 'Direct Taxes' levied on an individual economic unit, usually based on the individual's, household's, or company's income or wealth.

Examples would be:

a. For an individual or household.
A Poll Tax – mainly of historical significance only.
PAYE, Income Tax, Surtax, and Capital Gains Taxation and Estate Duty – the last two excluded from Table 24.
Wealth Tax – not used here but under consideration.

b. For a company.
Profits Tax, now superseded by Corporation Tax.

Indirect taxes are levied on a good or service. Who bears the burden of the tax* depends upon the market mechanism which may result in all the tax being absorbed by the seller, by the buyer or shared between them. There is no general rule.

Examples would be:

a. Purchase Tax and Excise Tax.
b. Rates.
c. Motor vehicle licences.
d. Social Security Contributions.†
e. Sales Taxes, Cascade Tax, and Value Added Tax – not in use in the UK although there is a lobby for the introduction of VAT partly associated with its use in the European Economic Community.

* This phrase is used to distinguish between who is affected by the tax in terms of paying a higher price, or having profits reduced, and the accounting method of remitting the tax to the government. Thus with purchase tax the cheque is drawn by the wholesaler but the effect of the tax depends on the market. (See Chapter Four.)

† This item (covering in the UK contributions for national insurance, National Health Service, industrial injuries, and the redundancy fund) is not always considered by the layman as a form of taxation. Their inclusion here follows normal practice in the technical literature which is based on the following reasoning. They are levied compulsorily and contribute to the difference between 'earnings' and 'take home pay', thus reducing disposable income. Although they give rise to negative taxation in the form of benefits, the funds are not run so as to break even, furthermore income may be redistributed by differences between the bases of payment and benefit. In practice, governments use changes in contributions as an arm of fiscal intervention.

TABLE 24

International Comparisons of Taxation 1967*
– as a percentage of GNP at Factor Cost

Country – ranked by Column 2 (1)	Total per cent (2)	Taxes on Income: Households (3)	Taxes on Income: Corporations (4)	Taxes on Expenditure (5)	Social Security Contributions: Total (6)	Social Security Contributions: Of which paid by employers (7)
Sweden	46·6	21·0	1·9	15·6	8·1	4·2
France	45·7	5·2	2·4	20·7	17·4	12·7
Norway	42·6	13·6	1·7	17·4	9·9	5·5
Austria	42·2	12·4	2·3	18·1	9·4	7·5
Netherlands	41·2	12·5	2·8	11·9	14·0	10·5
Germany	40·9	9·5	2·5	16·9	12·0	6·2
UK	37·7	11·8	3·0	17·4	5·5	2·8
Denmark	37·3	16·1	1·1	17·9	2·2	0·8
Belgium	35·8	8·2	2·4	15·4	9·8	6·8
Canada	34·8	9·2	4·2	17·6	3·8	2·1
Italy	34·8	5·8	2·0	14·9	12·1	n.a.
USA	30·8	10·9	4·5	9·7	5·7	2·9
Switzerland	23·6	8·5	2·3	7·5	5·3	1·8
Japan	20·3	4·5	3·9	8·0	3·9	2·4

– as a percentage of total taxes

Country – ranked as above	Taxes on Income: Households (8)	Taxes on Income: Corporations (9)	Taxes on Expenditure (10)	Social Security Contributions: Total (11)	Social Security Contributions: Of which paid by employers (12)
Sweden	45·0	4·1	33·4	17·5	9·0
France	11·3	5·3	45·2	38·2	27·8
Norway	31·8	4·0	41·0	23·2	12·8
Austria	29·3	5·5	42·8	22·4	17·7
Netherlands	30·3	6·8	28·9	34·0	25·5
Germany	23·3	6·1	41·3	29·3	15·1
UK	31·2	7·9	46·2	14·7	7·4
Denmark	43·0	3·0	48·1	5·9	2·2
Belgium	23·0	6·7	42·9	27·4	19·0
Canada	26·5	12·0	50·5	11·0	6·0
Italy	16·8	5·7	42·8	34·7	n.a.
USA	35·4	14·7	31·5	18·4	9·5
Switzerland	36·0	9·6	31·8	22·6	7·8
Japan	22·0	19·2	39·6	19·2	11·8

* Excluding taxes on capital, *eg*, Estate Duty, Capital Gains Tax.
Source: OECD's National Accounts Statistics. Adapted from Tables A and D, International Comparison of 'Taxes and Social Security Contributions', *Economic Trends*, No 187.

If we look at the percentage of GNP paid in total taxation (col. 2), we find little support for the commonly held belief that we stagger under an unusually high tax burden. Although the position of Switzerland as a European 'tax haven' is confirmed, we find ourselves ranked below France and Germany. That Sweden heads the list would not surprise most of us, but we suggest that the close proximity of a country with such a different reputation concerning social values as France would not be the usual view. Frenchmen may not pay their taxes, but those that they do pay take 8 per cent more of GNP than in the UK. We rank in an intermediate group, roughly mid-way between the USA and the leaders.

More meaningful analysis comes from a consideration of cols. 8–12, though it must be borne in mind that the percentages are proportions of total taxes, the relative size of which should be checked in col. 1. The first point to note is that only Denmark and Canada collect a smaller proportion of their taxes from Social Security contributions (col. 11). We should expect this result from our analysis of expenditure (page 183). Column 12 (and col. 7) show how the absolutely lower proportion and the 50/50 split between employees' and employers' contributions in the UK mean that taxes on labour borne by the employer here are low compared to France, the Netherlands, and Germany. In France all family allowance contributions are made by employers. Thus labour tends to be cheaper in the UK – although the inclusion of SET would affect the comparison. A widely-held belief is that the 'Welfare State' reduces the incentive to save. It may do this, but nevertheless in Germany, where Social Security contributions are just over twice the UK proportion (col. 6), personal savings are 2½ times the proportion of disposable income (12·6 per cent compared to 4·8 per cent).

One distinction between direct and indirect taxation which we have not yet mentioned is that from the viewpoint of the total tax paid by the individual; direct taxes can be designed to be 'progressive', *ie*, that those richer in terms of income and/or wealth can be made to pay a higher proportion of their riches than someone who is less well off. The nature of progressive direct taxation in the UK is discussed below.

On the other hand, purchase tax and excise tax rates paid are the same for all. While the rate will be the same, however, the total amount of tax paid will depend upon the pattern of consumption. The tax-paying performance of the chain-smoking, hard-drinking driver of a large car is proverbial. What is more important for comparisons between income groups is the type of commodity on which taxes are levied, so far as their importance in the budget of a 'poor' and 'rich' individual is concerned. In this sense, purchase tax and excise tax are 'regressive' in the UK, *so that a family with two children and an income of £16 per week pays twice the proportion of its income in this way than does one with an income of £60 per week.*

Bearing in mind this regressive effect, and our general belief that we are moving towards, or have reached, an egalitarian society, it is very significant that expenditure taxes here are a higher proportion of total taxation than in all other countries quoted, with the exceptions of Canada and Denmark, (col. 10). Another feature of our purchase and excise taxes is that they are narrowly based on a few commodities. Thus, taxes on tobacco, alcoholic drinks, and oil in the mid 1960s (pre SET) accounted for about half of total UK taxes on expenditure. Comparable proportions for Germany were $\frac{1}{4}$, France $\frac{1}{6}$, and the USA $\frac{1}{7}$.

The general picture which we see demonstrated in Table 24 is that, ignoring Japan, there are two major patterns of taxation and the UK system lies between them. There is a Continental European pattern of higher overall taxation, particularly resulting from high Social Security contributions. On the other hand, there is the USA, with rather lower taxation, more dependent on taxes on income. The major difference between the UK and USA for both the level and nature of taxation is not in terms of social security contributions (col. 6 or col. 11) but in the way in which firms here contribute a smaller proportion of the tax bill. (Col. 9 plus col. 12 gives the result: 15·3 per cent of total taxes UK, 24·2 per cent USA.)

By this time, the reader who believes that the high tax

burden is the major cause of our economic difficulties may be muttering 'but look at household's income tax'. This view of the high relative importance of direct taxation normally rests on two beliefs. Firstly, that income tax is the only tax the payment of which is noticeable and inescapable unless money income is deliberately reduced. Secondly, that in order to avoid high rates of income tax the labour force reduces its supply of effort. The second is unproved and the first clearly a misconception. All wage earners pay social security contributions and although we may shift other spending patterns in response to indirect tax changes* it is impossible to escape such taxes entirely. Indirect taxes may also have a disincentive effect – usually thought of in terms of direct taxation as discussed below – if, for example, basic consumption items are less taxed than luxuries.

Let us now turn to the alleged disincentive effect of direct taxes on the supply of effort. We assume that our specimen reader will not be mollified by col. 3 which shows that although we now move up the ranking we are still much in line with other countries except Japan, France, and Italy. After all, what is important is what he pays; this will depend under a progressive system greatly on his income. Surely the fact that high income earners may pay 19*s* 6*d* in the £ is important since these are likely to be the enterprising leaders of industrial society whose contributions to economic performance will be vital?

Overall percentages are thus not so important as the relationship between taxation and income levels. Furthermore, the average under a progressive system will be misleading since the marginal rate will always be higher than the average rate and it is the marginal tax/income ratio which the individual will consider.†

Certainly at the highest levels of income, marginal rates are higher in the UK than elsewhere. On the other hand, the

* And are meant to do so, the origin of high taxes on alcohol in the UK was a view in the 19th century of the social distress caused by drunkenness.

† We ignore another complication of the British system whereby a distinction is made between 'earned' and 'unearned' income.

average income tax and surtax paid as a percentage of income for those with incomes over £5,000 in 1966 was 50 per cent and this applied to less than 1 per cent of the population. The vast majority of income earners have standard rate as their marginal rate of taxation applying to part of their income. Thus we have a system which is progressive at low income levels, neutral over a wide range of taxes, and steeply progressive at high earnings.*

We now turn to the disincentive effect. This depends upon what taxpayers think they pay rather than on the actual payment. The distinction is important since it is our own experience, confirmed by survey evidence,[3] that most individuals overestimate their total tax bill and are incapable (if they pay Standard Rate) of calculating the extra tax which they will pay on a 'rise' of £1 per week. The answer is not 8*s* 3*d* but 6*s* 5*d*, due to the 2/9th Earned Income Relief allowance.

Faced with a group of management students we have often asked how many of them are working less hard as a result of the disincentive of personal taxation. This question highlights the practical difficulty of the argument, *viz* how is the supply of effort reduced. Three ways seem possible: less effort per hour, fewer hours per week, or an early retirement discouraged by it all. A further important point is whether the disincentive is from higher earnings at a given rate of tax or a reduction in effort as a result of a tax increase.

Looked at in this way, we feel that the results of several inquiries, which show little or no evidence for a disincentive effect in any of the forms mentioned, should seem reasonable. Few managers can greatly vary the length of their working week and both time and the intensity of effort are more likely to be associated with the job, their ability and

* If we include the regressive effects of indirect taxation we find that an individual earning £10 per week pays a higher average *total* tax rate than one earning £16 per week who in turn pays about the same proportion as the £50 per week earner. This result would be modified by 'negative taxes' – Social Security benefits – which however, depend not only on income but size of family etc.

ambition and their relationship with superiors, than with the level or rate of change in taxation.

Another way in which high marginal rates of income tax may cause a shift in the supply of effort follows from the lower marginal rate (which is, in fact, a constant and not progressive one) on capital gains. This, it is felt, shifts some potential high income earners into occupations, *eg*, in the financial markets of the City, where the rewards are less highly taxed capital gains. The economic implications of this shift, if it is an important one, depend on calculations about the marginal benefits to the economy of these two uses of labour.

We would agree that high tax rates may affect the supply of labour between various types of occupation, tending to favour those with high tax-free advantages in the form of untaxed benefits (*eg*, expense allowances, a car), social prestige and career certainty, (the Civil Service, academic research) and, of course, long holidays. In fact our colleagues concerned with human resources in management emphasize the importance of non-monetary inducements in the supply of effort, promotion etc. In many cases those who are in a position to vary their supply of effort are less likely to do so because of the importance to them of non-financial rewards. In particular, the tax advantages of being self-employed may tend to make occupations where this is comparatively easy, such as shopkeeping, relatively more attractive than being a wage earner paying by PAYE.

It is suggested that the manager's subjective view of his tax burden as high is due to miscalculation, high indirect rates on a small range of consumer goods, an unnecessary fear of the disincentive effect on the output of others, enhanced by the pleading of a small high-income group – to which if he is very unusual he may belong.

Two examples of the pervasive results of the belief in tax disincentives are the effect of taxation on overtime working and the 'brain drain'. The evidence on the former is that overtime working is associated with need rather than with tax paid. The common example of workers who, when offered evening or weekend work, decline because of the high marginal tax burden can be explained by the convenience of taxation as an excuse rather than having to explain to the foreman that at the margin the worker prefers leisure to income. The latter explanation may be an unpopular one to a manager faced with output targets to meet. The former is more plausible, even if based on a miscalculation.

Concerning the 'brain drain', a report in 1967 said that for most emigrants any tax differences if properly understood would be unlikely to sway a decision. This still leaves open the question of the effect of misconceptions about international tax differences – misconceptions which are perpetuated by the normal method of comparison in terms of direct income taxes for high income groups. One common source of error is to compare only the *direct* taxes paid by an executive earning, say, £3,000 *pa* in the UK and one earning the dollar equivalent in the USA. Apart from the fact that the official rate of exchange used may not reflect the real purchasing power of the two incomes and the unreality of measuring only part of the total tax burden of the individual, the fundamental reason for such comparisons being valueless is that £3,000 *pa* is a high income in the UK (and therefore highly taxed on a progressive system) but a comparatively low income in the USA.

Another feature of the UK tax system to be discussed is the nature and extent of changes over the last five years. Public expenditure has increased during this period and this, coupled with the government's decision to reduce the annual borrowing requirement from the private sector, has meant increased taxation. Whereas in theory inflation alone should cause a more than proportionate increase in receipts from progressive direct taxes, the lack of real progression in the UK system has slowed the increase. And of course some taxes, *eg*, rates, do not change at all with incomes.

Comparisons over time of total tax burdens are made difficult because payments and receipts of Social Security benefits depend so much on the composition of the family unit. However, it would seem that the burden on households has been reduced. Thus a family with an income of £21 per week (about the national average) retained 82 per cent in 1963 and 85 per cent in 1967. Similar figures for £60 per week earnings were 65 per cent and 67 per cent. The fact that there has not been a big increase in taxes paid by households is connected with new sources of revenue being found, *viz* SET, increased Social Security contributions by firms, and the larger revenue from corporation tax as opposed to the old company tax system. To put the total tax increase in perspective it would now seem to add between 2–3 per cent to the 37·7 per cent shown in Table 24, col. 2.

Finally we consider another aspect of the public sector, namely the extent to which it borrows from the private sector. In Chapter Three we have distinguished between the economic effects of savings and investment for the economy as a whole. In analysing National Income accounts any difference between the two will consist of new borrowing from or lending abroad. Briefly, the public sector has invested more than it has saved, resulting in borrowing from abroad (severely limited as a result of undertakings to IMF as a result of Balance of Payments problems) and borrowing from the private sector itself. Thus, nationalized industries cause only a borrowing requirement – one that is planned to decrease over the next few years as there is a lull in investment programmes. Central government has saved more than it has invested and the net borrowing requirements are very largely the result of heavy local government borrowing. It is important to distinguish between the real and money effects of this position. The real effect is a transfer of resources to the public sector, the results of which depend on the use to which they are put; the money effect inside the public sector is largely a result of decisions concerning the relative sizes of central and local government incomes and expenditures.

In conclusion, it should be remembered that the objectives of this chapter are to describe, measure, and discuss some economic effects of the size and nature of the public sector. We are not seeking to change attitudes to government intervention as such; only hoping to foster a more realistic and logical appreciation of the situation. We do not claim that, judged by any criteria, our public sector is perfect, still less do we believe that our tax system is effective – the strange results of personal direct taxation and the narrow range of goods subject to indirect taxes are strong evidence to the contrary. Our approach to the public sector has been limited to those fields which economists are particularly qualified to discuss. We are not competent to evaluate the bureaucracy, the short-sightedness of administrators and executives which undoubtedly exists in this, the single most important component of the economic environment.

Appendix: The Selective Employment Tax

SET was introduced in September 1966 at a time when the economy was on an inflationary part of the trade cycle and high priority was placed by the government on the objectives of growth coupled with solving the Balance of Payments problem. In its original form* it consisted of a tax based on the number, age, and sex of employees in an establishment (at rates of 25*s* per man, 12*s* 6*d* per woman, or boy under 18, and 8*s* per girl under 18, designed to be roughly proportional to earnings so as not to discriminate between sexes), coupled with rules which determined whether the tax, or tax plus premium, was repaid. Thus all employers pay at the same rates, although, of course, the burden depends upon the age and sex composition of the firm's labour force, but repayments discriminate between a 'neutral', a 'taxed', and a 'premium' class of activity. The neutral sector comprises: agriculture and extractive industry, transport, and government, all of which receive back, after a time-

* Subsequently the rates and repayments have been varied and a Regional Employment Premium superimposed – the current tax rates are 48*s* per man, 24*s* per woman or boy, and 16*s* per girl. Since 1968 the premium is paid only to firms in Development Areas on the basis of 37*s* 6*d* per man, 18*s* 9*d* per woman or boy, and 16*s* per girl. It is intended to collect the tax only from the 'taxed' sector when the new National Insurance Scheme is introduced in 1972.

lag, the amount which has been paid. The 'taxed' sector, which receives no rebate, consists of construction, distribution, banking, insurance, and other services. Finally the 'premium' sector, comprising manufacturing and related scientific research, received back its tax payment plus an additional 7*s* 6*d* per man, 3*s* 9*d* per woman and boy and 2*s* 6*d* per girl.

What was new was not the idea of a tax on labour – payroll taxes had been suggested for many years – but the unit on which the tax is based, the financing element and the discrimination between types of activity.

A novelty was using the establishment as the basis for taxation rather than the firm. An establishment normally consists of all premises at a particular address. It may be neither 'the factory' nor 'the firm': thus a firm with a factory in Maidenhead, another in Bournemouth, and a London head office, would have three establishments. The term is a statistical one, used for classifying output in various analyses, rather than a legal one. This has led to complications and anomalies in the operation of SET.

The unit on which the tax is based is of great importance since allocation to a sector was in terms of the Standard Industrial Classification, and 'premium' establishments only qualified as such as long as a majority of the employees at the establishment were counted as being engaged in manufacturing and allied scientific research.

By the financing element we mean the interest-free gap between payment and any repayment. Most taxes are paid in arrears – and this is one of the problems of 'fiscal policy'. In this case, even if all a firm's establishments are in the neutral sector, it is continually lending the public sector an amount which varies with its SET tax bill and the length of the repayment period. Thus, usual methods of estimating the impact of a tax by its annual yield are not accurate for SET, particularly since the cash-flow effect will depend upon current credit conditions.

Bearing in mind our assertion that a technique can only be judged by reference to its objectives, we will analyse

SET in terms of its likely objectives. 'Likely' because it is now difficult to judge whether all the objectives, listed 1–5 below, were held by the authors of the tax, still less to rank them in importance.

1 SET as a source of revenue. Initially its yield (tax less rebates) was estimated at £240 million per year. After the increase in rates (to make male 48*s*, female 24*s*) in July 1969 the expected annual yield was £606 million. These figures are small compared with total tax-revenues but equivalent to an additional 6*d* – 10*d* on standard rate for income tax. As a change in taxation there was a considerable disinflationary effect. If one accepts the need for a larger tax income, then the government's decision to do so by SET seems a reasonable technique rather than using politically unpopular increases in direct taxation, or purchase tax increases, which would have made Prices and Incomes Policy more difficult. Certainly those who wish to repeal SET are unlikely to be able to reduce expenditure by this amount and have to propose increases in existing taxes or new taxes to yield rather more than the actual yield ('rather more' because of the financing element). This means that SET is unlikely to disappear overnight. In practice the yield has been about $\frac{3}{4}$ of the estimate, largely because of absorbed SET lowering the profitability and thus tax revenue from some firms. As mentioned in Chapter Two there has also been a tendency for workers to become self-employed so as to avoid both SET and higher Social Security payments.

2 SET as a method of broadening the tax base. We have shown in this chapter how concentration on a narrow base for indirect taxes plus low Social Security contributions and high marginal rates of income tax are a feature of the UK tax system. This has encouraged suggestions for new forms of taxation, particularly since our system of purchase tax is inappropriate for the service sector. SET can be said to result from the government's rejection of VAT or a general sales tax – arguments about which are outside our scope. The administrative attractions of SET were ease of intro-

duction, since it used the existing National Insurance machinery and cheapness – government costs of collection being about 0·5 per cent.

3 SET as a method of changing the balance of taxation between goods and services. Prior to SET about 40 per cent on average of the price paid for goods by consumers (excluding distribution costs) was taxation – as opposed to 1 per cent of the price of a service. There is no scientific basis for the value judgement that taxes on goods and services should be equated, and certainly SET does not go far to redressing the balance. However, most countries do not find it necessary to have such a large distinction. Firstly, there are large reallocations in the price mechanism; thus laundries, launderettes, and home washing by machine are three ways of obtaining clean clothes but no taxes were paid for the first two methods whereas washing machines are taxed at $36\frac{2}{3}$ per cent. Secondly, in so far as UK indirect taxes are heavy on 'luxuries' but avoid foodstuffs, many service luxuries were exempt.

At the retail level SET does not simply discriminate in favour of services since many taxed services (*eg*, accountancy) form part of the costs of the manufacturing sector.

4 SET as a tax which would not harm the Balance of Payments. The rules of the General Agreement on Tariffs and Trades forbid rebates to exporters of the type of indirect taxes used in the UK. Thus, compared with such a tax increase, SET, by favouring manufacturers (who on average export 25 per cent of output) and penalizing services (8 per cent) would have a less harmful – and probably insignificant – effect.

5 SET as a method of initiating structural change in the economy. It is this objective, using the tax to promote growth, which has been most widely misunderstood. The view that the UK growth performance has been poor by international standards was discussed in Chapter Two as was the lack of any agreed reasons for the situation. One theory of economic growth (like all other theories, not accepted by all economists), is that of Professor Kaldor – at the time of

the introduction of SET an economic adviser to the government. Part of his explanation for growth differences was based on relating the growth rates of various countries to changes in the relative size of their Primary, Secondary, and Tertiary Sectors* and he found a correlation between overall growth and differences between such growth in the Secondary Sector. The explanation for this is disputed and complicated but rests on the 'Verdoorn Law' that fast rates of change in productivity are associated with fast increases in output. Kaldor then showed that the manufacturing sector was most likely to benefit in this way by the existence of economies of large-scale production. Thus increased productivity is associated with growth rather than stop–go 'shakeouts'.† One further, generally accepted, feature of the development of economies is needed – that over time the Tertiary Sector will grow in importance – and a theoretical basis for SET is completed.

To sum up: in Kaldor's view, Britain's slow rate of growth is a result of our having reached maturity – in the sense of following the US tendency for the Tertiary Sector to become the major source of GNP – at a lower level of income per head. On the other hand, fast-growing countries such as Japan and Italy have benefited from increased productivity from a growing Secondary Sector.

If we require faster growth it follows that one technique would be to hamper the development of the Tertiary Sector and seek to shift resources from there into the Secondary Sector. The Japanese and Italian solution of transfers from agriculture cannot work here due to the small relative size

* Primary Sector = agriculture and mining. Secondary Sector = manufacturing. Tertiary Sector = services. Their correspondence to 'neutral', 'premium', and 'taxed' sectors is obvious.

† Great weight has been attached recently to the economic costs, in terms of growth forgone, of 'stop–go' cycles. Such costs are the result not only of reduced investment but of the increased sophistication of firms in 'riding' the techniques in other ways, *eg*, by hoarding labour in times of 'stop' for future 'go' parts of the cycle. In so far as SET is seen as a policy associated with a theory which plays down the applicability of 'stop–go' policies it, possesses advantages to be offset against these costs.

of the Primary Sector. No value judgement that services are less 'necessary' than manufacturing is involved.

SET would, on this reasoning, shift labour from 'taxed' to 'premium' sectors partly by changing relative labour costs. In so far as the tax was passed on in increased prices, and the evidence is that this applies to no more than 2/5 of the tax, then the reduction in the 'taxed' sector would come from consumers reducing their purchases. The difference in total costs caused by SET between 'taxed' and 'premium' sectors will obviously vary from firm to firm, but an average is probably about 2 per cent. Thus, where SET has been completely absorbed by firms, some re-allocation of labour could be expected as a result of a shift of resources from less to more profitable sectors.

At this point we can introduce a widely known criticism of SET, that it operates the wrong way round, *ie*, if we wished to increase productivity, by encouraging the use of labour-saving machinery, the correct technique in the manufacturing sector would be to tax labour rather than provide a subsidy. Firstly, the whole argument for taxes on labour as a spur to productivity rest on dubious logical foundations. Secondly, Kaldor was not concerned with changing productivity at a given output but with growth by encouraging structural change. He saw the labour shortage in the manufacturing sector as a barrier to this type of change and proposed SET as a technique for overcoming the problem.

Having dealt with the nature and objectives of the tax, it now remains, before summing up, to consider some objections and problems not already mentioned.

The major objection – that the tax discriminates – may be based on three arguments:

a) That discrimination is wrong. This is a perfectly reasonable view to hold, but we must point out that all taxes discriminate, *eg*, between earned and unearned income and between the type of purchase for investment grants. By itself, this objection applies equally to all taxation.

b) That it is incorrect to treat broad sectors of industry as being homogeneous. Certainly the problems of, say, the construction industry and tourism may be very different. To some extent this can be compensated for in other ways, *eg*, when SET was introduced the construction industry was at the same time allowed to benefit from investment grants.* Broad categories, while increasing some possible anomalies, do reduce the 'borderline' disputes which have complicated the administration of the tax.

c) That the wrong sectors have been discriminated against. The theoretical basis for the three sectors has already been discussed. Another reason for objecting is naturally special pleading.

A more serious practical problem has been the assumption made that labour is sufficiently mobile for the transfer of labour resources to occur. Increased government assistance to labour mobility may have helped (see Chapter Five), but a cost of SET has probably been some increase in unemployment, particularly among part-time and older workers. SET may also have tended to reduce the size of the working population since the service sector is a large employer of married women, particularly on a part-time basis, who enter and leave the labour force according to the local availability of work (see Chapter Two).

The development of problem regions may have been hampered by SET which discriminated against areas where services are important (*eg*, the tourist industry of the South-west), and tended to adversely affect services in other regions where the lack of a growing service sector has been held to be an impediment to regional development. The Regional Employment Premium was designed to overcome this difficulty.

Although to the layman this appendix may have seemed an apologia for SET, its aim has been quite different. We

* This highlights a limitation of this sort of analysis of the effects on one tax. For the policy maker the tax system must be looked at as a whole.

have tried to consider the tax in the light of its objectives. So far as 1–4 are concerned SET has a strong case for consideration as a success, assuming that the case for other new types of taxation is not considered stronger. So far as 5 is concerned there is recent evidence that the transfer of resources is taking place, although ultimately an evaluation of the costs and benefits remains a value judgement.*

References

1 Peacock, A. J., and Wiseman, J., *op. cit.*

2 Buchanan, J. M., and Tullock, G., *Calculus of Consent*, University of Michigan Press, 1965, and Olsen, M., *Theory of Collective Action*, Harvard, 1965, contain some treatment of these two problems.

3 Brown, C. V., and Dawson, D. A., *Personal Taxation, Incentive and Tax Reform*, PEP Broadsheet, 1969.

Bibliography

Hill, T. P., 'Too Much Consumption', *National Westminster Bank Quarterly Review*, February 1969.

Peacock, A. T., and Wiseman, J., *The Growth of Public Expenditure in the United Kingdom*, Unwin University Books, 1967.

Sandford, C. T., *Economics of Public Finance*, Pergamon Press, 1969.

* Since this Chapter was written, the first part of an official inquiry into the economic effects of SET has been published (Reddaway, W. B., 'Effects of the Selective Employment Tax. First Report on the Distributive Trades', HMSO, 1970). The aim of the full report is to quantify the effects of SET compared to alternative taxation yielding the same revenue. In considering the economic implications of SET, Professor Reddaway attaches much greater importance to the tax as a revenue raiser, and less to the structural change and growth objective than we have done.

For the distributive trades as a whole he finds: (a) no evidence that the tax has been passed on by increased margins; (b) that there has been an increase in productivity by 1968 to the extent that 130,000 fewer workers have been required than would otherwise have been the case; (c) that there has been some reduction in service (*eg*, by the growth of self-service, supermarkets etc). He recognizes that it is difficult to separate the effects of SET from the abandonment of resale price maintenance. The report also contains suggestions for avoiding some of the anomalies in operating the tax.

CHAPTER SEVEN

Planning

Planning is simply the rational use of assets to achieve objectives. More precise definitions of types of planning are given later. At the moment no greater elaboration is required; however, it is worth emphasizing that using assets to achieve objectives is a process that takes place over time and involves using, or committing, assets now to meet current objectives at a future date.

Used in this way, 'planning' is a process which can be carried out by individuals and firms as well as government. We seek to control our consumption now in order to be able to go on holiday – 'planning a holiday' is concerned not only with where, when, and what to do when we get there, but with seeing that the resources will be available. Investment decisions by a firm are an example of part of a planning process. There is a certain amount of public hostility to planning in general; but, in our sense, we are all planners.

In fact, this use of the word may seem so wide as to include all economic activity and to destroy the often posed distinction between 'planning' and 'private enterprise'. 'Planning' and 'private enterprise' are not at opposite poles; what we hope to show in the rest of this part of the chapter is that a crucial distinction still remains. The second part of the chapter concerns itself with some available forms of planning while in the final section the British planning experience is analysed.

Private enterprise (or the price mechanism) in its 'pure' form consists of a series of institutions – firms, based on the private ownership of real and money assets – and a mechanism of prices which connects these producers with their markets. Thus firms are linked with their markets for outputs and inputs by prices. An increase in the wage rate will cause a firm to reduce the proportion of labour used in production. Where this is not possible, and where the increased cost is passed on in higher prices, there will be a tendency for fewer units of its output to be sold. The fundamental forces are those of supply and demand although the type of market structure, *eg*, competition or monopoly, will affect the price paid and received and the quantity bought and sold.

The mechanism of supply and demand operating through a market structure cannot itself have objectives. It may have characteristics, *eg*, the relationship between prices under conditions of monopoly or competition mentioned in earlier chapters. The institutions (firms) are assumed to be following their own self-interest, which is simplified to an objective of profit maximization in much of economic theory. What economists of the 19th century attempted to demonstrate was that the welfare of society, as a whole, was maximized *under certain conditions*. These economic theories did not consider profit maximization to be a good thing in itself; still less did they believe that every action taken by a firm to promote its objective was in society's interest. They considered that it was the *working of the system as a whole* which produced the desired result.

In contrast, planning inescapably requires the setting of objectives followed by a system to carry out the objectives. It should be clear that the adoption of a particular objective, *eg*, to discover a new method of production, carries with it the need to follow a new policy, *eg*, enlarge the firm's research facilities.

The conditions under which the welfare of society is maximized using the price mechanism are briefly:

1 That there is competition between firms to sell and between buyers to buy. This occurs when there are a large number of economic units involved and where no 'big fish' dominates 'minnows' in the pool.

2 That there should be a free market for commodities unhampered by outside control and free from agreements between sellers (or buyers).
3 That the market should be open in the sense that new firms would be able to compete without disadvantages, such as lack of access to new technology or money capital shortages.

How closely a particular market approaches the perfect mechanism and what practical significance should be attached to deviations from the pure form, have been extensively debated. It is enough for our purpose at the moment to reiterate that private enterprise has never perfectly fitted this model in the sense that there have always been industries monopolized or dominated by large firms.

The theory implied that to obtain the desired result it was necessary for individual firms to be 'price takers', *ie*, they merely respond to market forces. In practice, then and now, many firms are 'price makers' so that they can seek to control their price environment by advertising, marketing, and research policies, and collusion about prices and terms of trading.*

The second limitation in the classical analysis of the market mechanism is that it ignored 'externalities'. The distinction between private and social costs and benefits has already been made clear and the point will not be pursued any further.

The third problem associated with the actual market mechanism is that knowledge and ability are imperfect. Thus, firms may not have access to all available information concerning techniques of production, their markets etc. Furthermore, their information systems may be poor in that they have no way of knowing when they are actually best placed to fulfil their objective. Associated with the

* 'The United States, no differently from United States Steel, cannot rely on the market, without direction, to integrate its widely dispersed and complex operations in such a manner as more nearly to achieve its objectives.' *Private and Public Planning*, Chamberlain, N. W., McGraw-Hill, 1965, page 205.

assumption that knowledge was either perfect, or obtainable at zero cost, is the treatment of all factors of production as being perfectly mobile. Clearly, they are not; as was pointed out in Chapter Two.

The fourth qualification concerns the pace and direction of technical change. Classical analysis, since it dealt with a static situation, assumed a given state of technical knowledge and then went on to describe how resources would be allocated. More recently, it has become clear that technical change is both a way in which resources can be used and a powerful influence on the price mechanism. This has led to the views on market structure and innovation that we have already considered.

This has, so far, led us to a conclusion that the private enterprise system is not a perfect one, both in terms of the limitations of the original theory, and the 'performance gap' between theory and the real world.

Managers at some point in an economics course frequently say 'this is all right in theory but'; these qualifications, which are often popularly applied to planning proposals, usually occur in economics classes when 'perfect competition' is being discussed. It is as well to remember that the efficiency in allocation which follows from the application of the price mechanism in conditions of 'perfect competition' is the intellectual basis for the support of private enterprise. There is a great difference between the workings of private enterprise in theory and in practice – a similar comment applies to planning.

There is now a presumption that the price mechanism is not ideal. This does not mean that we should not continue to use it, but it does open the way for proposals for improving its performance* or allowing an alternative way of doing things, namely by planning.

Our next step is to demonstrate that the private enterprise system, excluding any government interference, is itself an important user of planning processes. It is here that our distinction between the *institutions* (firms) and the

* Such as those considered in Chapter Four and Chapter Five.

mechanism (prices) is important. For our purposes, the distinguishing characteristic of a firm is that it represents the part of a private enterprise system where resources are allocated by direction from above (*eg*, 'the Board', 'the Works Manager') and not by market transactions. The oldest large-scale examples of this type of organization are armies. It is no accident that the administrative structures which they have developed have been important in the development of management theory.

The firm can be considered as the grouping together, under single control, of a number of what would have otherwise been market transactions. This can be seen clearly if we consider the growth of a firm into a new market or stage of production. What has happened when, *eg*, a car manufacturer takes over a producer of pressed steel bodies is that a set of market transactions – the selling and buying of the components – is eliminated. The market mechanism shrinks as the scope of an institution organized to use assets to achieve objectives is increased. It has been pointed out how a planning unit expands as the importance of the market mechanism is diminished.[1] Why does the price mechanism continue to exist in wider markets while the coordinating unit of the firm is used in smaller markets? Why we do not find production controlled by one gigantic firm? This is explained by supposing that as a firm grows in size its costs of operation rise, until it is advantageous not to grow any further but to rely on the market mechanism. Private enterprise has always involved a considerable planning element and has, over time, had to face many of the problems associated with government planning.

The increase in size of the largest firms during the last 50 years means another distinction between public and private enterprise – the scale of operations – has been whittled away. General Motors commands greater resources than most of the world's governments. The objectives of a firm, as well as the structure of control, are very different from that of a government; but in terms of the directness of its ability to command resources a large firm is favourably placed compared to Western governments. It can direct labour, open and close plants in a way not available to them. Furthermore, it can vary the volume of resources which it uses in a way impossible for a government. To merge is an available strategy for firms, wars of conquest are abhorred for governments – although groupings such as the EEC, EFTA, and Comecon can be seen as alternatives.

What we have tried to show is that a common view, which considers 'planning' and 'private enterprise' as opposites, misrepresents reality. This is partly because countries provide us only with mixed examples of the two, but mainly because planning is a necessary part of any system of using resources to attain an objective, whether by an individual, a firm, or a government.

Why, therefore, is there widespread public antipathy to government planning in some Western countries (principally the UK and USA)? To be more accurate, the antipathy is towards planning in one's own country – 'a plan' is often a necessary condition before aid is granted by advanced to underdeveloped countries. A more uncomfortable example of this point is American insistence that reconstruction plans should be prepared by European countries who received 'Marshall Aid' after the Second World War.

Much opposition is generated from opposition to government intervention in general. This has already been discussed; planning is then seen as an effective government technique for extending and tightening control over resources. This attitude probably helped to perpetuate the short-term nature of government budgeting until quite recently. There was an air of 'spend the appropriation which you will receive this year and wait to see if any taxes will be raised next year'. Ironically, British governments came under considerable public pressure in the 1950s to plan the use of their resources. This was particularly the case with defence spending following the failure of 'Blue Streak' and other advanced weapons.

Much opposition comes from those who claim that planning has been tried and failed, but some would say, so has the price mechanism. This will be discussed further in the British context in the third section of this chapter.

Lastly, there is opposition to particular forms of planning. These views will often be irrelevant to the discussion of possible practical planning techniques for the UK; although on a political level, it is often claimed that any form of planning is the thin end of a bureaucratic and/or socialist wedge.

Three political events in the 20th century which had enormous economic implications have been the Russian Revolution and the two world wars. The former instituted a political system which led to a planning system by 1928. The USSR did not return to peacetime conditions until 1921 and operated 'state capitalism', even associated with some denationalization, until 1928. To a great extent the political attitudes of Western democracies associated together the political and economic aspects of the post-1917 regime. This worked in the reverse direction when the First Five-Year Plan from 1928 seemed to some observers to be the only solution to Western problems of massive unemployment in the 1930s. Certainly Russia did escape this aspect of the private enterprise system, as it then existed, but popular interest fell away with the realization that Keynesian stabilization policies could avert the problem, with perhaps increased intervention, but no commitment to long-run planning. The essence of Keynesian analysis was its concern with short-term symptoms; Keynes himself pointed out that 'in the long run we are all dead'. This horizon is a sensible one for measures designed to correct cyclical fluctuations, such as those described in Chapter Four, but it differs from the planning horizon. After all, governments are aware that any plans for future action, even next week, are communicated to a population, a proportion of which will have died by the time that they are implemented. For a nation the future is unlimited; governments as a result have much longer time horizons than individuals or firms.

For reasons already discussed, the world wars caused increased government intervention of a planning type. The objective was clear and the inapplicability of the price mechanism in directing resources was accepted. But in the UK physical controls continued for some years after the Second World War (rationing was finally abolished in 1954). These wartime and post-war restrictions, while accepted as being caused by the war, became associated almost as a *result* of the planning controls and are part of the consciousness of those now older than 30. It is often said that labour relations are complicated by the 'fear of unemployment', which has a special meaning to those who were socially aware in the 1920s and 1930s – even though it is assumed that such a situation (where for nearly 20 years the national unemployment rate only rarely dropped below 10 per cent) could not re-occur. There may well be a parallel here with the stance of those whose attitudes to planning were shaped by wartime and post-war 'austerity' conditions. Because, of course, in political terms the debates about extensions and contractions of planning in the UK are never involved with a form of planning which has similar objectives or controls to those of 20 years ago. If, however, we refrain for the moment from considering the practical issues and consider the theoretical balance between the two systems we can demonstrate the differences, and similarities, most clearly by comparing the two extreme positions, *viz* central planning and the market mechanism.

We will start with listing some of the deficiencies of central planning from a 'Western democratic' point of view.

1 That central planning involves state ownership. Economic theory has nothing to say about the form of asset ownership. Voters may find the idea of state ownership attractive or repugnant; economics can help to provide insights into the results of any pattern of ownership. What needs to be made clear, however, is that whereas private ownership runs contrary to the objectives of socialism, and central planning would be impossible, in practice, without a considerable amount of public ownership, socialist planners do not reject prices as a mechanism. Thus, in the

USSR there has always been a small private ownership sector,* some goods have been subject to the price mechanism† and, since 1956, labour has completely been freed to shift between jobs in response to wage signals from the labour market.

But, in the main, Soviet planning, while it may use planned price differentials to influence the pattern of demand, does not use price as a mechanism to clear the market. An overpriced commodity in a Western country would remain unsold until the price fell (*eg*, by means of a 'sale'). In the USSR prices are not flexible – although the planners and producers may be made aware of dissatisfaction with the price or quality of a consumer good by entries in a 'complaints book' kept at the point of sale for this purpose and by mounting stocks of unwanted consumer goods.

Thus, in theory, private enterprise market equilibrium is brought about by price and output changes, whereas only the latter are used in central planning. Once again, in reality the systems may produce similar results; a feature of markets dominated by a small number of firms in the private enterprise system is that prices are changed infrequently. Thus, changes in demand for cars (*eg*, caused by variations in HP regulations) result in changes in the output of Ford Cortinas rather than in a variation of their price. The price pressure exists, and may be reflected in the discounts which dealers are prepared to offer; but only rarely is the list price changed.

In the price mechanism, prices refer both to outputs and inputs. A superficial discussion of central planning often concentrates on the outputs ('black market' prices for Western nylon shirts), whereas the input prices have caused far more difficulties in Russia. This is because Marxist economic theory developed the 'labour theory of value' which held that only capitalist systems produced prices for the other factors of production, land, and capital. But the function of a price in a mechanism for allocating resources

* Chiefly in the craft industries and agriculture.
† Some agricultural produce.

is to measure its value in terms of scarcity. As in all other economies, land and capital are scarce goods in Russia. Misallocation was bound to follow so long as an economic unit (*eg*, a factory) was not charged for *all* factors used by the planning mechanism.

2 That it is impossible in practice to plan centrally for a large advanced country. If we mean by this a system whereby decisions relating down to the level of the firm are the results of edicts from one planning authority, then Russian planners would agree. They might point out that to use such a system in General Motors or Shell would be equally impracticable and that, like large private enterprises, they have developed techniques to overcome the problem. These techniques are concerned, firstly, with ways in which the key variables for the whole economy can be analysed; secondly, with communication systems to allow the required knowledge to flow up the hierarchy and, thirdly, with the devolution of decision taking.

3 Opposition to the objectives of the planners. This again is essentially a political matter. Even the results of decisions about objectives, such as the well-known Stalinist concentration of resources on capital rather than consumer goods which, it is held, has resulted in a faster growth rate, may be objected to for the same social reasons that one may object to growth in a Western country.

A Digression on Soviet Planning

To attempt to cover the theory, or practice, of a Russian plan is outside the scope of this book, but it is essential to outline some important elements at this stage. Firstly, there are several plans, each with a different time horizon, in operation at any moment; the basic document is, however, the annual plan to which our comments relate in this digression. Each annual plan starts with knowledge of the results of the previous period. This information provides the basis and linked to it are growth rates (which may be inherent in the current Five-Year Plan). The assumption is that all resources will be used (in contrast to indicative planning which has to accept some level of unemployment). The analysis starts with the calculation of 'leading links' – certain key relationships, *eg*, energy production, which are known to be troublesome. Then for each of 130 basic raw materials a 'balance' is produced, *ie*, the source of the inputs and use of outputs of an industry are tabulated (planning is for a balance since the aim is to use all resources). It is recognized that many crucial constraints can be thrown up by looking at 'horizontal' implications of changes in production. This is done by input-output analysis. Since Russian planning is in physical units, not money value, the example outlined below is possible in, for example, French planning, but would be incomplete in Russia. The columns could not be added up. Russian economists pioneered input-output analysis, greatly developed in the USA by Leontief. This is a technique whereby the whole economy is divided into sectors (*eg*, agriculture). The contribution to the total from each sector is therefore available and *so is its contribution to intermediate stages of production and its requirements of raw materials.*

		INPUT			
		Agriculture	Manufacturing	Household Consumption	Total Output
OUTPUT	Agriculture	20	15	65	100
	Manufacturing	20			
	Households	20			

The diagram above should be viewed as the top left-hand corner of a chess board, the rows and columns of which would be extended in practice to cover, for example, numerous subdivisions of the manufacturing sector and could take into account imports and exports. Our example shows that for agriculture the total of 100 units comprises (reading across the row) 20 units of production transferred elsewhere within the agricultural sector (*eg*, production of oats for cattle food), 15 units transferred to manufacturing (*eg*, barley for distilling), and 65 units of produce for household consumption. This output is obtained (reading down the column) from the 20 units of input from the same sector, already mentioned, plus 20 units from manufacturing (*eg*, farm machinery) and 20 units from households as owners of resources (*eg*, wages, rents, profits, and interest). Such input-output tables are now available for many countries.*

If we assume that inputs are used in constant *proportions* as output changes, then the implications of, *eg*, an extra 10 units of consumer produce being produced, can be traced as 10 per cent increases, firstly in inputs bought within the agricultural sector, secondly on demand for agricultural machinery and thirdly for resources. But these, in a full table, could be seen to have still further implications for resource use.

This emphasizes the point that equilibrium in any economy can be considered as a vast series of simultaneous equations, so that changing the values in one part of the system by switching resources (whether planned or the result of market forces) has widespread implications elsewhere. In a private enterprise economy the solution of the equations is found by a process of iteration, *ie*, approximations gradually converging on the correct solution. Thus an increase in demand for potatoes will lead to an increase in their production, plus an increase in the quantity of farm machinery and farm labour required. The reaction of the farming sector to the increase in demand may be excessive at first, but will ultimately settle down (as will the other sectors affected) to reflect the new level of demand. Planning, on the other hand, can attempt to predict the results by a process of analysis.

* For a brief, non-technical, description of the technique and its application to underdeveloped countries, see Leontief, W. W., 'The Structure of Development' in *Technology and Economic Development*, A 'Scientific American' Book, Penguin Books, 1965.

The basis of the theoretical situation was quite clear 40 years ago and at that time it was sufficient to point to the vast amount of calculation necessary under a centrally planned system to demonstrate its impracticability compared to the spreading of the burden under private enterprise. This argument is no longer true, analogue computers can solve the equations quickly (but the problem of obtaining the information still remains, see below). Let us suppose that this is possible, what advantages can result? Firstly, we could work through the process of iteration in abstract with our only costs those of the planning mechanism, whereas the whole system in the price mechanism uses real resources and, therefore, has real effects on living standards and patterns of consumption.

Secondly, we can use the linear programming technique.* Whereas input-output analysis is concerned with demonstrating mutually compatible levels of output from sectors (valuable for the economy as a whole), linear programming is concerned with the best use of resources (a valuable technique for both the economy and the firm). Furthermore, it has been shown that, assuming no differences in objectives, a perfectly formulated linear programme would result in the same processes and outputs as those in the economist's model of perfect competition. The qualification is of great importance. As we have seen there is no single objective for a free enterprise economy. In a centrally planned economy resources can be allocated rationally to meet objectives which must be given to the planners. Except in a completely totalitarian situation, when the problem does not arise, a great practical difficulty is to identify consumers' preferences. Central planning tends to provide solutions in terms of inputs rather than outputs.

Thus, theoretically, an economy plan using linear programming could produce the result of the 'ideal' in free

* This was discovered by Kantorovich in 1939 (and suppressed for ideological reasons). It was independently discovered in the USA in 1947, since when it has been widely applied by large private enterprise companies.

enterprise. A key difference, however, is that linear programming is an exercise which throws up values of resources used in production to provide the signals in the system. These values ('shadow prices') represent the marginal costs of using resources and would be the same as money prices under a system of Paretian optimality. These 'shadow prices' perform the function of prices in the (perfectly competitive) price mechanism where a firm accepts prevailing market prices. But, whereas a price in the price mechanism is 'real', in the sense that it actually occurs, it is possible to run various programmes on a computer corresponding to various objectives and then select the actual one to be used.

'Shadow prices' are, of course, a feature of planning by large corporations; they provide a basis for internal (*eg*, inter-divisional) 'transfer prices' within the planning unit. Such prices are necessary when, for example, a firm manufactures an item which is then retailed by another part of the organization. They may be of great importance outside the firm when the goods are manufactured in one country and transferred to another. 'Shadow prices' are not the only basis for such 'transfer prices', which have been the subject of wide debate among accountants. Other widely used valuations are 'cost of production' and 'market price'.

To return to our list of criticisms of planning:

4 All this does not mean that planning can at present produce the results of a perfectly competitive market mechanism. But it does emphasize that meaningful comparisons can only be made on the basis of ideal *v* ideal, or actual *v* actual mechanisms. The major shortfall in the centrally planned system is the result of the problem of coordinating dispersed knowledge. This again has had to be faced by large private enterprise units and the solution adopted by both sets of planners is to set up systems which filter all knowledge so as to allow only the fundamental parts of the

whole to be passed up the organization.* Decentralization for this reason (which has been greatly extended in the USSR since 1965) extends the area of competence of decision takers at the lower levels. Thus, factory managers in the USSR should have increased powers very similar to their opposite numbers in Britain. Some decentralization may, therefore, be a practical necessity, as well as being politically desirable for those who believe in the importance of preserving some local control over local interests. (In practice this has still to take place as State bureaucrats have hung on to their authority.)

In terms of the market mechanism the parallel is between perfect competition and oligopoly (an industry which comprises a small number of large firms). The latter may have all sorts of practical advantages but cannot give the same 'ideal' pattern of resource use that follows from competition. Similarly with planning, centralized planning offers the hope of the ideal, whereas a system of giving budgets to sectors (*eg*, agriculture, transport) and leaving the allocation of resources within each sector to its planning authority can never produce 'ideal' results. It may produce the only practicable system of operation – as with firms who allocate resources between parts of the enterprise by means of budgets.

5 This gives rise to the fifth planning problem, namely how specifically shall targets be laid down for decision takers at lower levels, bearing in mind that their objectives may not be the same as their superiors. This point is discussed again in Chapter Eight. Clearly the objectives of a foreman are likely to differ from a sales manager who, in turn, has a different motivation from the managing director. Target specification has given rise to some of the legendary stories of Russian planning; such as the nail-producing factory which, when given a target specified in weight of output, produced very large nails, when given a target of number of nails

* Such as the well-known technique of 'management by exception' or 'leave it alone until there is a variance between results and the target'.

produced very small ones and when asked to produce 'nails over 2″ long' produced all one size, $2\frac{1}{16}$″.

The problem of reconciling differences in objectives is common to all large organizations, public or private, profit maximizing or charitable. It is unrealistic to expect a track worker at Fords to have the same objective as his managing director, just as the objectives of the nail factory manager differed from the planning authority. Again our readers will be familiar with the problem and the various attempts to solve it in their environment, *eg*, by setting targets for sectors of the business which are consistent with the overall objectives. Our readers will also be aware that no ideal solution has been found by large firms. Managerial control systems may produce 'cost centres' but this often implies cost minimization as the objective of such sub-groups. While this may be a reasonable 'rule of thumb' in the short-run it will be sheer chance if it coincides over a longer period with the likely objectives (*ie*, profitability, which involves revenue) of the whole organization.

6 The last item in our catalogue of objections is the problem of error. It is held that central planning suffers from error in the sense of differences between planned output and achieved results.* The problem of errors resulting from central planning is often stated as being firstly, that no one is gifted with foresight and secondly, that the errors will be bigger and in one direction. The first part of the case is true and applies to all of us – but it no more destroys the case for government planning than it does for corporate planning – or for that matter planning by individuals. Nor is it true that the size of the uncertainty to be faced is greater for a

* We exclude two other types of error: 'wrong' objectives, a political matter considered in 1 above, and slowness to react to changed conditions. We know of no evidence to enable a comparison to be made between central planning and private enterprise on the latter point. Although in many fields Russian innovation has lagged behind (say) the USA, this has been partly because of a disinterest in marketing innovation (styles of motor vehicles) and partly because Russia has been able to select and use from the innovations produced in the Western world. See Chapter Five for an example of reaction time of a major US industry.

government than an individual. For most people the decision to buy a house involves incalculable risks which would deter the professional.

The problem of size of error is a real one, but we cannot fall back on a belief that if there are a large number of decision units their errors will cancel out. Statistically, this assumes no connexion between their decisions whereas in practice they are all likely to react in the same direction to a market situation. The history of the machine-tool industry is full of such examples.

Of course there is considerable evidence of bad national planning if the sole criterion for judgement is the closeness of the actual results to the plan. Any manager who has operated a budgeting system will be aware, however, that the important thing is not the size of the difference but how it arose. If the difference is due to unforeseen events then as well as sacking the planner we must remember to alter the plan target.

What we hope to have shown in this part of the chapter is that the two extreme cases of Perfect Competition and Perfect Computation[2] face the same problems and would solve them to give the same answers – *in theory and assuming the same objectives.* The only theoretical difference is that in planning all costs and benefits can be included, whereas in private enterprise social costs and benefits are ignored or undervalued.* The great operational, as opposed to political or social, advantage of the price mechanism is that it decentralizes decision taking and thus reduces the high cost of information collection and communication necessary under a planning régime. Planning may tend to stifle initiative, although there is no conclusive evidence one way or the other on the point. Central planning needs administrators and probably requires more than the less centralized struc-

* Strictly speaking this holds good when we consider a world of no technical change. Some writers have claimed that, considered dynamically, it is advantageous to exclude the social cost-benefit results of new investment as this would tend to lower its 'total profitability' and thus discourage acts which are both responses to and initiators of technical change.

tures of private enterprise. We may prefer the 'price mechanism' operated by privately owned business units for social and political grounds, but to the economist each system has economic costs and benefits. Certainly both systems in practice fall short of their theoretical possibilities.

We now turn to considering the techniques of planning which have been followed in Western countries and which have provided the basis for the approach to planning in Great Britain which forms the final part of this chapter.

One form of planning common to all 'advanced Western' countries can be dealt with quite quickly, namely environmental control such as 'town planning' and 'regional planning'. Although this is probably the most common context in which the term 'planner' is used, it is not our main concern, which is with planning applied to the whole economy. Countries engage in restricting and directing land use because of an implicit acceptance that private enterprise concerns itself with private and not social effects. The basis for regional planning was dealt with in Chapter Two, the evidence for the need to plan the urban and rural environment will be known to our readers in terms of local examples of 'urban blight', 'ribbon development' etc. Some aspects of this type of planning are the results of the State's responsibility for providing communal services such as roads and public housing. In this field of planning, governments have a poor record measured by the current aspirations of society. Thus the motorway programme has been criticized on the grounds of quantity, quality, and siting of the new roads, with the current hysterical controversy over the 'London Motorway Box' a topical example.

It is now generally accepted that government planning is necessary and inevitable in this type of activity. The questions involved, apart from the correctness of the decision in the light of its objectives, are: whether the planning authority should be central or local, if local of what type, and whether the planning should be carried through to actual construction or be limited to 'outline planning' (*eg*, land use).

Overall planning of the type with which we are mainly concerned has been carried out in Britain since 1962. As will be mentioned later, the impetus came from industry and politicians, as well as economists and planners. The motives were a combination of disillusionment with the effects of short-term government intervention ('stop–go policies') coupled with admiration for the results of *French planning*.

French planning is often termed 'indicative planning', in contrast to 'imperative planning' of the Russian type, and in its pure form could be described as a process whereby the authorities indicate economic objectives for a number of years ahead (in this case four) and elaborate the actions necessary to achieve them. French planning has always gone beyond this by having central government regulations and inducements directing both the public and private sectors of the economy towards achieving the objectives of the plan. For this reason it is more accurately termed 'flexible planning' or 'economic programming'. We will continue to use the term 'indicative planning' however, since our concern is ultimately with British planning and, as we shall see, planning here has been more indicative than its French counterpart in practice, though probably not so much in intention.

What are the advantages claimed for indicative planning – apart from those already mentioned for planning in general and the political suitability of this form for an economy with a basically private enterprise system coexisting with an important government sector?

At the time, great emphasis was placed on the value of indicative planning as a means of obtaining faster economic growth. In one sense this was inevitable since 1961 was the dawn of 'growth mania' in this country and almost any piece of economic policy had to have growth as a promised reward to be worth consideration. Planning has continued to be associated with growth and growth rates, which emphasizes a difference from other techniques (*eg*, welfare services, taxation) which have been associated with the reallocation of resources.

Growth was to be assisted through indicative planning by

'raising the sights' of businessmen who would be made aware of the implications for their firm of the achievement of a predetermined growth rate in the economy as a whole. In its simplest form this could be predicted by the application of the principle of self-realizing expectations. One example would be if a sufficient number of investors on the Stock Exchange formed the view that a particular share will rise in price and act rationally on that belief (*ie*, by purchasing the share), then their expectations will be broadly vindicated. For example, if sufficient businessmen act on the belief that GNP will rise by 25 per cent over the next five years then there is a good chance that such a growth rate will be achieved. If the forecasts on which the expectations are based are wrong, then the results of incorrectly based decisions may be harmful – although it is always possible to be right for the wrong reasons.

This is a very simple view of the effect of expectations, and supporters of the case that 'demand constraints' limit growth would expect the plan to increase confidence, not only by 'selling' the growth rate but by identifying the structural economic problems of achieving it (*eg*, by input/output analysis) and incorporating techniques to overcome them. Much of the case for indicative planning *and growth* rests on the acceptance of 'demand led' theories of which a special case is 'export-led growth'. This is argued to have been important for countries in the EEC, particularly for Italy and Western Germany. But in so far as indicative planning fosters growth by raising expectations there is clearly a problem regarding export-led growth, which requires not only changed expectations but changed overseas demand. This could result from: the growth of world trade in the prime exports of a country; an increase in its competitiveness due to lowered costs or innovation; exchange-rate changes or joining the EEC (likely to be mainly a once and for all effect). But exhortation is not going to help.

We have now considered three different growth theories in this book. The case for investment as the important causal factor was mentioned in Chapter Two and Kaldor's view of the importance of structural shifts considered in the Appendix to Chapter Five. In addition, the historical importance of technical progress has been stated several times. Conceivably all could be important, but recent work[3] places emphasis on supply problems (*eg*, investment, shifts in the labour force) rather than the demand constraints which were more academically fashionable in 1961. Of course it is quite possible that indicative planning may act favourably on some of the supply problems. Thus, increased investment may result from firms now having access to better information regarding the future requirements of other sectors of industry.

The second positive benefit claimed for this form of planning is the dampening of trade-cycle fluctuations. This could come from any ability which the plan has to forecast the structural results of technical change, the advantage of operating on a longer time horizon for dampening fluctuations in investment or the demand for durable consumer goods. Great importance has been given by some writers to the problems faced by the price mechanism in the current economic environment where much investment can only be carried out on a large scale and over a long period. As a result, fluctuations in such industries are likely to be large and will take a long time to adjust to an equilibrium level (if the term is at all appropriate). Businessmen may all hold back from investment through the disincentive effect of uncertainty over a long time period or, alternatively, over-investment and surplus capacity may occur. The increased importance in total demand of the consumption of durable items (*eg*, cars, refrigerators, and washing machines), produced by capital-intensive techniques and where demand consists of new entrants to the market plus re-ordering, has introduced another unstable element.* Better information and better forecasting resulting from the indicative planning process could thus help 'balanced growth'. But, as always when we consider growth, there is a contradictory view, *ie*, that 'un-

* The treatment of consumption as stable and investment as volatile in the Keynesian analysis in Chapter Three.

balanced growth' is faster. This latter view holds that it is the uneven rate of advance, by firms and industries, and the opportunities of high profits from 'being right' and from innovation, that provide the sharpest spur to technical progress – and thus faster growth.

Another way of stating the case for indicative planning as a provider of growth and stability is to compare the alternatives. If these are considered to be 'stop–go' policies which (it is claimed) restrict investment and cause their own cyclical fluctuations then planning may be desirable by comparison. This would seem to have been the popular view here in 1961.

Before recounting the planner's progress in Britain (and seeking to explain the Slough of Despond into which he has ventured) we will look at the model on which he based his programme.

The impetus for the First French Plan in 1946 came from the urgent popular requirement of rapidly rising living standards after the post-war devastation and public political acceptance of planning as a means of achieving this. In addition, there was the requirement of American administrators of Marshall Aid that forecasts should be made of its expected results.

French plans (we are now in the era of the Fifth) run for four years. They are projections of the state of the economy at the end of the fourth year in the light of planned objectives. These have developed from being simply growth rates to 'growth plus control of inflation'; social goals, such as the improvement of the educational system and the absorption of Algerian refugees have also been included. There has been a consistent long-term commitment to forms of Prices and Incomes Policy, Regional Policy, and the maintenance of public sector investment, which contrasts with the fluctuating involvement of British governments.

In addition to the political and economic interest of the Cabinet, planning is administered by a Commissariat du Plan (a small Civil Service department of specialists with overall responsibility for the preparation, submission, and implementation of a plan) and Modernization Committees.

The latter comprise unpaid part-time members representing owners and managements, labour, the Civil Service, and independent experts.*

These two special types of planning body have no direct powers; planning is, after all, indicative or flexible and decisions are reached by consultation and agreement rather than by imposed authority or voting procedures. Nevertheless, the government has various techniques for encouraging industries to follow the plan's objectives.

Firstly, there are direct controls on building, on location in the Paris region, and on the construction of oil refineries. Secondly, since most medium-term credit is State controlled by the Credit National and permission is usually necessary to make a share or debenture issue, the authorities can, and do, ensure that external finance is only available for firms who conform to the plan. Thirdly, and most importantly, a wide range of inducements is available, such as subsidies, tax rebates, price controls, and contracts (under which a firm may undertake, *eg*, to manufacture a specific product or undertake specific research in return for, *eg*, a tax waiver or a price advantage).

The technique of planning can be described briefly as follows. Firstly, forecasts are made of key components of national product in four years' time, based on various growth rates. Since the Third Plan, this has involved the use of an input–output matrix of 28 sectors. It must be emphasized that French planning is not only flexible but partial. Obviously consumer demand is not planned; although forecasts of its components are made at this stage. Not all industries are included in the planned forecasts and the groupings of those that are included into 28 sectors would be considered hopelessly crude by Russian standards – and compares with the 81-sector matrix in use by analysts in the USA.

Simulations are carried out using various growth rates

* In the rough proportions 40 per cent, 10 per cent, 25 per cent and 25 per cent. The similarities and contrasts with British Economic Development Committees are discussed below.

from which one is selected by the Cabinet, on advice from the Commissariat du Plan; this is usually the highest reasonable rate of growth taking into account the social objectives given to the plan by the Cabinet.

The next stage is, apart from the feasibility of the objectives, critical for the success of a plan. The Modernization Committees prepare for their industries targets of such key components as output, manpower, productivity, investment, and foreign trade. These are inspected for overall consistency by the Commissariat du Plan, after which the Committees prepared detailed recommendations as to how the plan can be carried out – down to firm level in some cases. Two points need to be made here. Firstly, the process just described involves an enormous amount of consultation, not merely inside a Modernization Committee but with the Commissariat du Plan. Secondly, whereas most MCs are 'vertical' (*ie*, they cover a group of industries with subcommittees for individual industries) there are five which are 'horizontal'. These cover: the general economy and finance, manpower, research, productivity, and regional development. Their function is to view certain results of vertical committees in a detailed way to ensure consistency with the more aggregated approach of the Commissariat du Plan.

Finally, the plan is presented for parliamentary approval, after which it becomes, at least during the time of President de Gaulle, an 'ardent obligation'.[4] In terms of the length of the overall description this may seem to imply a perfunctory quality to parliamentary involvement in planning. This reflects a real political problem, namely how can a parliament effectively influence planning when it is asked for a yes/no decision? Tinkering with one or two components is not allowed as this would upset the internal consistency of the plan as a whole. Here the economist bows out, leaving the debate, which is important in Britain as well as France, to the politician – theorist or practitioner.

What have been the results of planning in France? One obvious test is to compare planned and actual results. In the case of the Fourth Plan (1962–65) the target was fulfilled exactly for Gross Domestic Product, while planned investment was exceeded by 5 per cent. While, other things being equal, the successful attainment of targets is desirable we would suggest that accuracy is neither the point of indicative planning nor a measure of its success. It is not the point, which is basically the effectiveness of the plan as a bundle of medium-term policies; it is not a measurement of success since perfection is unobtainable in practice and differences may be the result of unforeseen contingencies. What is more serious about the Fourth Plan is the variations from target in particular industries, *eg*, oil – 27 per cent, machine tools – 12 per cent, coal – 16 per cent. These discrepancies cast doubt on the inner coherence of the plan and the accuracy of the input-output analysis.

There is great controversy concerning the connexion between planning and the performance of the French economy. In the late 1950s and early 1960s the French growth rate classed that country as one of the 'fast growers' with Italy and Japan. Steady growth was achieved in a fluctuating political environment, apparently only at the sacrifice of exchange-rate stability. Subsequent performance has not been as steady – although still higher than that of the UK. The recent political and economic problems connected with rapid inflation and devaluation have caused observers to downgrade, in their view, the success of French economic policies and thus French planning – and the German growth rate, achieved in a 'non-planning' country, has been held up as an example.

A final point that needs emphasizing about planning in France is that it takes place in a social climate quite different from that of the UK. This is not, of course, a point for or against the French system as such, but is material when we consider the adoption of indicative planning in the UK. The first example of social differences is that of the civil servant as a planner. In France the State, since Napoleonic times, has intervened in directing the economy and this direction has been exercised by civil servants rather than by parliamentary laws. French planners are accustomed to being 'political' (they form part of the Cabinet) and to exercising initiative and partiality* in order to attain planning objectives. English civil servants are only prepared to operate a policy. The French are much readier to accept direct intervention, in contrast to the usual peacetime climate of anti-leadership here.

The implications of this difference in climate are far reaching and may be fundamental. Like any other project, planning is helped enormously if those involved in the execution are committed to the course of action being undertaken. The government and the planners are usually involved in this way in central planning. In indicative planning great use is made of consultation with private industry who may well be sceptical of planning; what is then required is commitment by those executing the plan. But we are not suggesting that civil servants ought to be planners.

Business representation on Modernization Committees predominantly represents large firms. In addition there is reason to believe that Committees may meet unofficially as Trade Associations – which could be illegal under British law.

The British Planning Experience

This section of the chapter concerns itself with the form and impact of planning for the whole economy in Britain. It traces the development of planning institutions and plans and finally considers some by-products of the national planning mechanism.

Indicative planning in Britain may be said to have started with the formation of the National Economic Development Council in 1961. This was a direct result of a report of the Council on Prices, Productivity, and Incomes.

Indirectly, there had been a favourable climate for indicative planning among businessmen (an FBI Conference

* This does not imply corruption. The fact that such an interpretation comes to the reader's mind shows the depth of the British tradition of Civil Service impartiality. Where this attitude to intervention is not followed, as in the case of the IRC, it results in public opposition.

in 1960, a conference organized by the NIESR in 1961), the public ('growth mania' explaining the French results as having been the result of planning), and politicians (Mr Macmillan had written, in the 1930's, two books advocating planning).

The NEDC comprises representatives of the central authority (Cabinet members and civil servants), private industry, and the trade unions in equal proportions, together with members who are independent, and others who represent the nationalized industries.* Its status has been described as 'under the aegis of but not in the administration'.[4] Its composition is an example of a peculiarly British approach to national economic problems in the 1960s; namely that there are three interest groups to be represented, *viz* the government, the private sector as firms, and the trade unions, and that meaningful criteria of the public interest can result from the authorities receiving advice resulting from the consultation of such a non-executive body.† While the association of independence and lack of executive power is reasonable enough, it has led to the situation in which the NEDC and the National Economic Development Office have not had access to the most recent government information, statistics etc. This was particularly true in the pre-1964 period when it suffered particularly from being neither part of the government nor completely independent.

In addition to its general responsibilities for identifying an objective (in terms of growth) and shaping a planning policy, the NEDC was required to consider: obstacles to growth, methods of increasing efficiency, and whether the best use was being made of resources. It was backed with the National Economic Development Office, a secretariat of about 100 specialists seconded from other fields.

In early 1963 the NEDC produced the first plan‡ based on

* Membership has varied over time. At present the total is 24, with the Prime Minister as chairman.

† *Eg*, the Joint Statement of Intent on Prices, Productivity, and Incomes, which initiated our present type of incomes policy and was signed by representatives of the three interest groups.

‡ As distinct from the First National Plan.

an industrial inquiry asking firms and industries for forecasts consistent with 4 per cent *pa* growth rate of GDP over the period 1961–6. As a result of consultation it was decided that 4 per cent was feasible. The plan resulted in a progress report (1964) examining the growth actually achieved and initiating a further inquiry among firms concerning their growth experience. All this seems to have had little real impact, but the Council's report on policies and techniques for attaining the growth target has been extremely influential. This report, 'Conditions Favourable to Faster Growth' (CFTFG, April 1963), indicated possible action in the following areas: education and training (The Industrial Training Act 1964); labour mobility and redundancy (The Redundancy Payments Act and earnings-related unemployment pay); regional policy (Industrial Development Act 1966); the Balance of Payments and taxation (SET, discussions about VAT), and prices and incomes (the National Board for Prices and Incomes). Action along the lines indicated was urged as being ways in which faster growth could be achieved. The influential nature of the report is shown by subsequent action indicated in brackets.

In 1964, as a result of NEDC pressure, Economic Development Committees were instituted. These were not the first British attempt to imitate the French Modernization Committees, since there had been an abortive try with Development Councils in 1947. Time had brought a change in the attitudes of most businessmen with the result that they were welcomed – and have continued to flourish. Their objectives are: firstly, to consider the economic performance, prospects, and plans of their industries and to assess progress in the light of the national growth objectives; secondly, to give this information to the NEDC, and thirdly to consider ways of improving the economic performance, competitiveness, and efficiency of their own industry. Once again, their composition is 'tripartite' as in the NEDC; the chairman is always a businessman drawn from outside the industry.

At present there are 21 EDCs covering about two-thirds of the private sector. With one exception (concerned with 'the

Movement of Exports') they are 'vertical' in that they represent an industry (*eg*, building, chocolate and sugar confectionery and, rather strangely, the Post Office). Thus they have never been concerned with the general 'horizontal' areas of competence of some Modernization Committees, although they have spawned specialized committees dealing with problems common to several industries, such as 'The UK Committee for the Simplification of International Trade Procedures'. EDCs, through their processes of consultation, can hope to reduce the subjective nature of the answers given by firms to the questionnaires which form such an important part of British planning.

By October 1964 we had: the NEDC, which had already some experience in formulating growth targets and examining obstacles to their attainment, plus EDCs which could relate these objectives down to the industry level as well as providing the information required for a forecast. At that moment the Department of Economic Affairs was created with overall responsibility for planning. Most of the NEDC was transferred to the new department, the NEDC being left with a much restricted field of operation confined to: firstly, acting as a channel of communication to, and between, EDC's; secondly, issuing specialized papers, organizing National Productivity Conferences and the like, and thirdly with formulating the overall 'norm' of permitted incomes and prices increases to be applied in detail by the National Board for Prices and Incomes.

Viewed pessimistically, the NEDC can be called a mere 'talking shop'. Its members from industry can be dismissed as being unrepresentative of business managers in general who have little time to attend conferences etc. Trade unionists have felt that planning is 'lad's work'. On the other hand, the government, the CBI, and the TUC attach considerable importance to the forum which it provides for debating economic policies, and its decisions regarding feasible growth rates (and thus permitted increases in prices and incomes) carry considerable weight. An example of this is the TUC's threat towards the end of 1968 to leave the NEDC during a dispute over growth rates. In this case the TUC favoured a faster rate of expansion than the rest of the Council would allow and was clearly aware of the implications for various methods of government intervention of choosing a particular growth rate.

The DEA produced the First National Plan in September 1965. 'The Plan' was based on a target of GDP 25 per cent higher in 1970 than 1964. It contained an analysis of how the increased GDP was to be allocated (*eg*, between personal consumption and investment) and six areas in which specified action was to be taken to help the Plan's success (these were virtually the same as in 'CFTFG' above).

Three parts of the Plan were expected to be particularly difficult to realize. One was the existence of a 'manpower gap' in 1970, in the form of a 400,000 difference between the demand and supply of labour – this has not materialized and resulted from errors in the compilation of the forecasts. Another was that the growth rate in any year (3·8 per cent) might be adversely affected by short-term government intervention designed to stabilize the economy and help the Balance of Payments. Lastly, both the growth rate itself and the necessary international payments results depended on a growth of exports of 5¼ per cent *pa*, compared to an achieved 3 per cent. This rate was not obtained and it was the crisis measures of June 1966, in response to severe Balance of Payments problems, which caused the Plan to be abandoned. Any 'blame' does not wholly lie at the door of exporters, however, since imports, international flows of capital, and the seamen's strike contributed largely to the 1966 situation.

Much has been written about the failure of the Plan, largely by those hostile to planning who see it as a failure of planning. What is remarkable is the absence of committed planners arguing that 1966 demonstrated the need for different targets. In fact the Plan, by 1966, had few supporters. We will examine why, classified by problems of objectives, of procedure, and of fundamental techniques.

To set a 'correct' target is one of the most difficult

problems in all types of planning, whether we consider a firm (*eg*, budgets), central planning, or indicative planning. A simple projection into the future of past achievements (3 per cent in 1962 and 1963) is unambitious. One alternative is a forecast based on the trend of growth. The Plan was more ambitious than either in that it relied on predictions of the effects of planning policies. To restate this in terms of budgeting in a firm, the choice was between, *eg*, sales last year, sales increasing at the rate achieved in the last few years, or sales based on past growth rates plus estimated effects of new management policies. In the end, the 25 per cent 1964–1970 increase seems to have been selected as much for political reasons as for its reality. Perhaps this merely reflected official euphoria with the 'raised expectations' effect of indicative planning, based on a belief that constraints on growth were on the demand side.

The planning procedure has also been heavily criticized. The whole operation was carried out in a great hurry – the DEA wanted to produce results quickly and there was a political commitment to early action. The information given to DEA planners (largely from NEDO) consisted of a growth target of GDP and certain other assumptions– such as a 3·5 per cent *pa* productivity increase and an 8·9 per cent growth in the world market for manufacturing exports. In addition, they had information regarding the forecasts of industries which resulted from prior NEDC inquiries. They then asked some 50 industries, approached via an EDC if one existed, if not via the Trade Association, for forecasts based on the 25 per cent GDP target of: output, investment, manpower, requirements etc. Where the replies proved mutually inconsistent those industries affected were asked to rethink and provide consistent estimates. Similar information was supplied by nationalized industries and ministries produced forecasts of the effect of policies (*eg*, labour mobility).

This procedure, while superficially similar to that in France, demonstrates the fundamentally inadequate approach to the Plan. Although use was made of the 'Cambridge Model' of input–output, the approach was one of iteration by asking for revised estimates from industries rather than the use of a tested input–output model for the economy by which the implications of various growth rates, assumptions of exports etc could have been tested and action then taken on the industries involved in bottlenecks. The use of input-output analysis in forecasting was untried in this country – as was, of course, the whole planning procedure. Experience in other countries shows that a necessary condition for a successful plan (in the sense of one for which the target is attained) is the existence of an earlier plan and model into which a correct appreciation of the present position can be fed.

Lastly, we must remember the wide enthusiasm for the notion of a plan, which made technical criticism difficult and which wholly obscured the important principle that the nation was being required to undertake actions regarding the use of resources (and involving, of course, social costs as well as benefits), for a growth target which had not been publicly debated. But then, growth was everything and demand constraints were officially considered to be the only important barriers to its attainment. There is much in the criticism that the First Plan was a series of existing policies tacked on to a statistical projection of what would happen if GDP increased by 25 per cent over the planning period.

The Plan was abandoned in July 1966 and by some this was seen to be the end of British planning. The DEA continued the planning process however, and, in fact, early in 1967 a CBI conference requested that more realistic, less detailed planning should continue.

Somewhat diffidently, the second attempt was launched in February 1969 as 'The Task Ahead'. Its sub-title, 'Economic Assessment To 1972' emphasizes the official view that this is 'a planning document not a plan' and is 'part of the planning process'.

The objectives are now stated as being:

1 The achievement of a substantial and continuing surplus in the Balance of Payments.

2 A steady improvement in the competitive efficiency of the economy.
3 An improvement in the regional balance of the economy.

Achievement of these, largely unquantifiable, objectives will lead to growth. This, however, is not stated as a single growth rate but a wedge (to use the document's own terminology) of results following from growth at 3¼ per cent *pa* 1968–72, or a better (4 per cent), or a worse (3 per cent), result. At the present time other forecasters see the 3¼ per cent rate as being feasible, although the planned annual growth in incomes of 2·4 per cent may be difficult to achieve, given the typical 3 per cent national growth rate during the last ten years.

The discussion of targets in 'The Task Ahead' attaches prime importance to the achievement and maintenance of a Balance of Payments surplus, as this could be the growth constraint. The basic 3¼ per cent GDP rate is based on a 3 per cent growth in productive potential (*ie*, output resulting from increases in the quantity and quality of factors of production, assuming the same level of utilization) plus the effect of increased exports. These have been rising at a 3 per cent annual rate; the basic growth target of 3¼ per cent requires them to grow at 5¾ per cent (comprising the achieved 3 per cent, an extra 2–3 per cent resulting from devaluation and a further 1 per cent from exports to fast growing markets). A slower growth of exports could clearly have two results for the Task; it could directly lower achievable growth and, if resulting in a Balance of Payments problem, could lead to other intervention which would damage the growth rate.

The planning procedure has also been changed. Emphasis is placed on the DEA's responsibility for the plan, but the importance of direct government consultation with the CBI and TUC is stressed in addition to the tripartite forum provided by the NEDC. Whereas, ultimately, all EDCs will be asked to examine the implications of the plan, in the first instance more urgent and detailed inquiries are going on with seven EDC's, parts of the public sector (energy and steel), and certain key industries not covered by EDCs (*eg*, ships and aircraft).

Greater use has been made of input-output analysis. This has been hampered because the intersectoral relationships are out of date, being based on 1954 figures, and the sectors themselves do not always correspond with the EDCs who are expected to examine the implications of the plan.

We are not in a position, at this stage, to evaluate the prospects of this further attempt at planning. It would seem that the private sector, as well as the government, remains interested in the value to all parties of better information about the likely progress of the economy over a medium time period. The switch in emphasis is clearly to a continuous process of consultation, aimed at obtaining benefits from better information, rather than increased expectations.

There are two minimal reasons why this sort of planning is likely to continue in Britain. Firstly, it offers the hope of an alternative to 'stop–go' policies, thus providing a more propitious background for private sector planning. Secondly, the considerable size of the public sector (see Chapter Six), which operates with different objectives and a longer time horizon than private enterprise, means that there will be continual pressure for accurate forecasting for the whole economy against which the government sector can design its investment policy.

National planning and national plans are only part of the adventures in the 1960s. There have been at least three important by-products *viz* regional planning, the industrial aspects of the EDCs, and Prices and Incomes Policy. The last named we have largely ignored,* but the other two are worth describing further.

Regional planning in Britain has, at present, three objectives: physical planning, remedying regional imbalance, and as a part of comprehensive national planning. Up to the

* Partly for reasons of space but also because of likely changes in the near future concerning the techniques and institutions involved. See page 247 for some comments concerning productivity and prices and incomes.

early 1960s, action concerned only the first. Thus the Greater London Plan, 1946, was concerned with problems of urban renewal and growth. This arm of regional planning has continued to be important, *eg*, The South-east Study, 1964, and parts of studies by the new planning regions. The CFTFG report of 1963 advocated a regional planning structure to enable the second and third objectives to be brought about. It was held that a regional planning programme would make national planning easier, and would allow higher growth targets to be attained by making increased use of resources and reducing inflationary pressure in prosperous areas.

As a result of this official attitude, strengthened by the effects of the 1962–3 winter recession in problem areas, eleven planning regions were set up in 1964. These comprised eight regions in England, plus Wales, Scotland, and Northern Ireland. Regional boundaries were drawn, to some extent, as a result of existing regional sentiment and administrative convenience. The choice of large areas reflected dissatisfaction with using small problem areas as the basis of policy. Clearly the choice of unit in such planning is vital; there is, however, no single economic decision rule to use in such a situation. Two alternatives would be: to choose a region in terms of its commercial links with a conurbation, or to define a region to include areas facing similar problems. The choice of the regional unit is of great importance when financial benefits are made available which discriminate between regions or when the central government allocates resources between regions. A good example of this has been the recent decision to help 'grey areas' – largely zones on the edges of Development Areas which shared the latters' problems without obtaining their benefits.

A Regional Economic Council and a Regional Economic Planning Board was set up for each region. Each Council has approximately two dozen members, drawn from local businessmen and trade unionists, local authorities, academics etc. Its powers are advisory and non-executive and its objectives lie in producing a long-term strategy for the region, concerned with the best use of resources. A Board comprises civil servants ordinarily situated in the region and is designed to coordinate the policies of their departments which have regional implications.

Councils for all regions have now produced their strategies embodied in studies (*eg*, 'A strategy for the South-east' 1967). Where the region is prosperous, these have tended to concentrate on physical problems (*eg*, overspill), where the region is a problem one the emphasis is on removing constraints on growth. None, however, could be called regional plans. In some cases this is partly a result of the lack of regional consciousness (the nearest to a plan so far has been the study for Scotland). More generally there is a problem that neither Councils nor Boards have any executive authority.*

The major problem, however, is in meeting the third objective of regional planning – the link with a national plan. Firstly, this has been made difficult where an attempt was made (as in the case of Scotland) by the abandonment of the National Plan itself. More fundamentally, however, the necessary information is just not available. We have no 'Regional National Income' figures (although a start has been made). The 'Regional Balance of Payments' is unknown. This is important both to calculate the multiplier effects of an increase in purchasing power in a region and to trace the impact of growth in a region elsewhere. Thus the construction of a car plant in Scotland may have considerable effects in the Midlands from where many of the components will be obtained.

Proper regional planning involves input-output analysis. As we have already mentioned, no up-to-date national model exists, still less regional versions. (It would be wrong to

* The only regional executive authority in Britain is the (Scottish) Highlands and Islands Development Board. Set up in 1965 this has a small budget and powers to physically undertake development. It appears to produce worthwhile results, but faces opposition on the grounds that policy decisions are taken and implemented without any electoral control, except, indirectly, via the overriding responsibility of the Secretary of State for Scotland.

assume that the inter-relationships in, say, South Wales were the same as for the United Kingdom.) What is really required to meet 'the planner's dream' is a sort of three dimensional input-output table with regions as the third dimension. This has, so far, not proved a practicable proposition anywhere – and would, of course, vary according to the groupings on the third axis (*ie*, the definition of a region).

So far as the second objective (regional imbalance) is concerned, much has already been said in Chapter Two. Once more we are short of the information regarding the costs and benefits of, *eg*, decisions to bring about re-location of plants.

The inter-relationships between national and regional planning are complex, since each can be said to depend on the other. Fundamentally, however, it is the responsibility of central government to allocate national resources between regions (*pace* Scottish Nationalists). This will involve arbitration between Regional Economic Planning Councils and between the central and regional authorities. One administrative problem is that the government is faced, not only with 'plans' from Boards, but with pressure from elected local authorities and local Members of Parliament. It would appear, however, although this is not very clear, that apart from the studies which have been produced by the regional planning framework, most continuing pressure comes from the older, established bodies.*

Where the objective is to reduce regional imbalance by re-location of firms and by providing resources for the regional infra-structure, then the central government's role is vital. Thus a prosperous region may require IDCs to re-allocate industrial development away from its congested conurbations, while the central authority may wish to give IDCs for re-location out of the region into a Development Area.

In conclusion, the present regional planning mechanism, for reasons outlined above, has, and can have, little impact on the first two objectives. It has been more valuable in the

* The North-east Development Council is an exception to this.

older field of regional planning concerned with physical problems of urban renewal, communal services, transport etc. Administratively we appear to be in a 'halfway house' waiting for the results of the new structure of local government and susceptible to pressure for elected regional authorities. The so-called regional plans still tend to become 'a chronicle of regional aspirations'.[5]

Our second by-product of planning is the system of Economic Development Committees which was set up in 1964 and has since been extended. EDCs have three main functions: as providers of planning information; as carriers of planning down to the level at which most action must be taken, and as consultative bodies which can look for weaknesses in their industries and consider the effects of government policies. We have already mentioned the present scope of EDCs and their relationship to national planning. Since British planning has always emphasized that the success of a plan depends upon action taken at the industry and firm level, and since there is no compulsion to carrying out planned objectives,* the EDCs have always been considered important in a second sense as providing the tripartite consultation which can lead to action. There has not been a study which measures the results obtained by an EDC, still less the cumulative effects of all of them. They have, however, consistently been rated as important by the authorities and have continued to be supported by management and unions.

Our purpose here, therefore, cannot be to analyse their effectiveness but to consider some of the possible undesirable effects of such an organization. What follows concerns the work relating to output. EDCs also consider existing government policy by which they are affected, complain, praise, and suggest new policies. For example, one result is that trade unions have agreed to support new working methods.

Much of the consultative work of EDCs has been

* Unlike France, there have not been any special techniques apart from those associated with policies developed for short-run intervention. See Chapter Four.

concerned with helping the Balance of Payments. Originally the emphasis was on promoting import substitution. Thus, the Mechanical Engineering EDC held a conference to discuss implications for the Industry of the NEDO Report 'Imported Manufactures: An inquiry into competitiveness' which stressed that the fitness of purpose of an imported commodity and its delivery date may be more important than price. The Machine Tool EDC, as part of a 24-point 'Action Programme', examined imports to see whether some types of frequently imported machines could be produced in Britain as a result of a concentrated and co-operative effort by firms.

More recently, the emphasis has shifted to the promotion of exports. The Chairman of NEDC pointed out that the usual argument that a healthy home market was important as a base for exports can be reversed to read a 'healthy Balance of Payments is a necessary condition for a truly flourishing home market'.* Various schemes have been suggested for promoting export effort in industries where there are many small non-exporting firms, *eg*, 'pick-a-back' where a large firm handles a smaller one's products as an agent. Many EDCs have discussed and implemented schemes for rationalization, reduction in the product range in order to obtain economies of scale, and the exchange of information concerning investment.

This type of discussion formed part of the agendas of the Trade Associations, many of which went on to discuss prices, terms of trading etc, prior to the 1956 legislation (see Chapter Four). We are not saying that EDCs are the 20th-century version of the Trade Associations developed in the 19th century. Many Trade Associations still exist (the Mechanical Engineering EDC covered 50 in 1965) having purged themselves, where necessary, of the practices barred in 1956. EDCs represent unions and third parties as well as firms. Possible benefits from better information and reduced uncertainty were outlined in Chapter Four.

* Letter from Brown, G., quoted in DEA Progress Report, July 1965, page 5.

But, in France, MCs have provided the institutional framework for strengthening Trade Associations – albeit in a more favourable climate. Certainly EDCs were finding their work affected by Restrictive Trade Practices legislation and obtained partial release from this in the 1968 Act.

The danger follows from the fact that Britain 20 years ago was riddled by agreements between firms regarding prices, outputs etc, to a much greater extent than any other 'advanced Western' country. Most were stopped a dozen years ago – but the habits of mind which they bred in some industries can only be broken down by time. EDCs may prolong the memory.

Secondly, EDCs (unless, ironically, they make use of a Trade Association) reflect the large firms of their industry. These may be the most advanced managerially, technically etc. But reports of agreements to cut down imports by the rationalization of the product range can cause unease. Firstly, there are well-established advantages to international trade. This is clear when we think, as we eat a banana, – should we persevere with attempts to grow them here? Clearly not; nor would there be support for attempts to diminish public consumption of this fruit. What about English versus imported apples? The discussion is now more 'marginal' (and complicated by the uncertain supply situation of fruit) but, unless you are a farmer, most would support imports in principle and do so in practice by buying. International trade continues on the basis of imports and exports. We benefit from imports as well as exports. To return to machine tools, it is at least worth asking the question whether resources are better diverted to 'filling import gaps'* rather than, *eg*, concentrating on cutting delivery dates on exports. The answer may be to do both, but this involves extra national resources being devoted to this industry.

* Especially if production takes place behind a tariff wall. The writers have experience of the advantage to be gained from a firm producing a machine to compete with an imported one (an EDC was not involved). As a result an importer cannot claim remission of any duty paid – this is only possible if there is no British equivalent.

Lastly, there is the point that an effective organization which represents some members of an industry will, over time, inevitably hamper structural change resulting from the emergence of new and small firms – which it does not represent.

Conclusion

This chapter has tried to show that planning is used in both the private enterprise and public sectors of an economy. The objectives are, of course, different but many of the techniques used, and the problems faced, are common to the manager and the government planner.

After looking at two well-developed forms of national planning – in France and in the USSR – we have considered the planning experience of the UK. Whatever may have been the results up to now in this country, we hope to have shown that the decision to plan is neither purely political, nor stupid and irrational. On the other hand, planning is not a panacea which solves all national economic problems. But some forms of planning will remain a feature of the manager's environment – particularly within the company for which he works.

References

1 Coase, R. H., 'The Nature of the Firm', *Economica*, November 1937.
2 Wiles, P., 'Imperfect Competition and Decentralized Planning', *Economics of Planning*, Vol 4, No 1.
3 Denison, E. F., *Why Growth Rates Differ*, *op. cit.*
4 Oxenfeldt, A. R., and Holubnychy, V., *op. cit.* page 205.
5 McCrone, G., *op. cit.*, page 243.

Bibliography

Barker, T. S., and Lecomber, J. R. C., 'Economic Planning for 1972: an appraisal of "The Task Ahead",' *Political and Economic Planning*, 1969.
Chamberlain, N. W., *Private and Public Planning*, McGraw-Hill, 1965.

Department of Economic Affairs, *The Task Ahead. Economic Assessment to* 1972, HMSO, 1969.

McCrone, G., *Regional Policy in Britain*, Unwin University Books, 1969.

Oxenfeldt, A. R., and Holubnychy, V., *Economic Systems in Action*, Holt, Rinehart, and Winston, 1965.

Polanyi, G., *Planning in Britain*, Institute of Economic Affairs, 1967.

Shonfield, A., *Modern Capitalism*, Oxford University Press Paperback, 1969.

CHAPTER EIGHT

Economic Efficiency

An improvement in efficiency is widely held in 'Westernized' countries to be a desirable type of change. This chapter concerns itself with efficiency in relation to various economic units, namely, the individual considered as a unit of labour input, the firm, the industry, the government, and the economy. It also deals with some of the problems involved in this type of measurement.

We must start by examining the two quite distinct ways in which the word 'efficiency' is used. Firstly, there is 'efficiency' in the physical sciences, used to mean a ratio between inputs and outputs. An example is the conversion of gallons of oil to British Thermal Units of heat produced by a central-heating boiler. Secondly, there is 'efficiency' in terms of the ability to attain an objective. Examples are the efficiency of a transport system in moving goods between places, a lunar rocket in reaching the moon, and the price mechanism in allocating goods and services. This second use of the term will be called *performance*. In addition, *efficiency* (in the first sense) may be measured absolutely or relatively; thus an oil-fired burner may be, by its design, more efficient than another, but both may be less efficient than some theoretical or practical standard of excellence.

The distinction between *efficiency* and *performance* will be maintained throughout this chapter, although of course the two concepts may give similar results when applied to a particular case, *eg*, an efficient typist. But this is not necessarily so. Thus one car engine may be more efficient than another in so far as it has a better petrol/distance ratio but has a worse performance, *eg*, a lower top speed. A further problem of performance is that there may be more than one objective. Thus a price system may be asked to *allocate* resources and promote their *efficient* use. Another point is that great complications arise when there is more than one output, *eg*, gas and coke, pork or bacon.

Efficiency and performance will both give the same results where the standard against which performance is measured is a target input-output measure. The importance of the distinction is that, whereas performance is usually a meaningful and measureable concept in the social sciences (but see page 258 below), efficiency very often is not. This is because, in the physical sciences, inputs and outputs are in real measures and in some cases the same measure (*eg*, BTU) may be used for input and output, whereas in the social sciences the two usually consist of quite different items (*eg*, pig-swill and sides of bacon) which can only be compared by using money.

But money, by the standards of a physical scientist, is a poor measuring rod* because of changes in its value over time and difficulties of converting one money form into another. Furthermore, when we are comparing the quantity × price of different items, prices may vary for reasons which may not be under the control of the economic unit being considered. It is possible in real terms to calculate a pig-swill/bacon level of efficiency and compare it with other pig-swill/bacon conversion processes. This could also be done for pig-swill/pork. But to compare the use of pig-swill for pork or bacon production involves a knowledge of the prices of the products. Given such knowledge a choice could be made with, say, the objective of maximizing the revenue obtained from a given quantity of pig-swill. But this would be a measurement of *performance*. The *efficiency* of the conversion could not be measured because the prices of outputs may be partly determined by monopolistic elements in the market, *ie*; the performance of the price mechanism is also involved.

The next problem which faces the social scientist measuring

* An example of relative 'efficiency', in the second sense of performance.

efficiency is the problem of multiple inputs and outputs. Furthermore, the proportions may vary between, say, inputs not merely because a different technique is being used, but because most techniques allow inputs to be used in varying proportions. The firm will vary these proportions as the relative prices of inputs change. Thus, in steel production, some plants are designed to accept greater variations in pig/scrap proportions than others. The proportions actually used within the technical constraints will depend on the relative prices of iron and scrap. The ability to vary the inputs and take advantage of market variations is an aspect of the performance of the plant so that efficiency for any given combination of inputs and outputs may be sacrificed for flexibility in the choice of the particular combination used. This may be very valuable where the relative prices of inputs (*eg*, scrap and pig iron) or the demand for outputs jointly produced (*eg*, petrol and oils from a refinery) vary.

To sum up, efficiency is a concept taken over from the physical sciences which has its usefulness for our purposes limited by problems of measuring inputs and outputs, both of which may be multiple and variable for a process in money terms. On the other hand, to use performance as a measure requires an agreed objective.

We will now consider efficiency and performance for each of the economic units mentioned at the start of the chapter.

1 *The individual*

An individual may seek to improve his own level of performance by 'trying harder' or for example, by attending a course on management. To the individual the worth of such a course will be the difference between the benefits received from the course as opposed to the alternative ways of using the time and money which he has had to forgo. Generally it seems more useful to look upon this example as being a method of increasing the quality of a factor of production, akin to the use of fertilizers on a field.

A popular measure of efficiency for the individual is that

of '*productivity*', defined as being the ratio between hours of work and output for a particular task. Such measurements are widely used within a firm, sometimes as the basis of the payment system, for making comparisons using the same type or different types of labour between industries, and for national and international comparisons. Productivity has become an important criterion used in government intervention. Thus the National Board for Prices and Incomes has stated that it sees as an objective for approved wage agreements that they should relate to productivity rather than comparability.* Our case is that productivity, as a term measuring a type of efficiency, is often misused, so that in many cases the measurement is applied in an unrealistic and misleading way.

By 'productivity' is normally meant labour productivity; clearly the concept could be applied to any other factor of production such as land or capital. But any productive process uses *more than one* factor of production and to measure output in terms of only one input may be seriously misleading. An example, familiar to accountants, would be the uselessness of choosing between two methods of production on the sole basis of which involved the higher labour productivity, *eg*, measured by output per man-hour. If each involved very different capital requirements, then a rational choice should include capital productivity as well. In practice, of course, costs per unit of output would be compared and not (labour) productivity. Another well-known problem, which illustrates the point, is the difficulty of relating earnings to productivity in wage agreements, because the productivity of some workers is unmeasureable.

* Despite the critical treatment which follows of productivity as a measure, to relate wage increase to productivity changes may be a reasonable 'rule of thumb' way of seeking to restrict increases in the price level by affecting the level of wages. This is quite different from implying that we are using a scientific measuring standard, or claiming that wages 'ought to' be related to output. The latter point is an opinion. Economic theory can show how (in certain circumstances) a firm seeking maximum profit will be helped towards that objective if it relates the value of a worker's output to his remuneration. This is not the same thing as saying that such a relationship *ought* to exist.

If productivity is used as a basis for payment, to what extent should earnings rise when output per hour increases as a result of using extra capital equipment in the process? Techniques have been developed to produce satisfactory answers to this question, but our point is that they can only be reasonably accurate inside an individual firm with a tailor-made agreement and when there is no great change in the techniques of production. To make use of production measurements for inter-firm or international comparisons of efficiency is to misuse the measurement. Inside a firm the cost of other factors may be ignored, or treated as being constant in order to simplify the calculation or because, over a short period of time, other variations are unimportant.

Internationally this is an unwarranted assumption because the proportions in which factors are used in a process depend on their relative prices and these differ widely internationally. Thus a few years ago at the time of the Chinese 'Great Leap Forward' a newspaper photograph showed an earth dam being built using long lines of Chinese labourers equipped with baskets to carry earth to the top of the site. This process was considered to be 'primitive'. Perhaps it was, but it was certainly a lower cost solution to the problem, given the relative shortage of capital and abundance of labour in China (which in a free price mechanism would be reflected in their price) compared with the UK. By extension, a similar dam built in the USA would use relatively more capital and less labour than in the comparatively cheap labour economy of Britain. Thus (labour) productivity is no help in comparing the efficiency of, say, steelworkers in the USA and UK. Of course, higher productivity in the UK steel industry may be considered desirable, but the economic environments of each country make it inevitable that there will be differences as the proportions of factors of production possessed by the economy. Thus in comparing British and American agriculture we should expect to find that, in terms of labour productivity, the USA would be ranked first, but using land productivity, output per acre, the UK would come off best. This is, in fact, what happens.

This leaves us with productivity, apart from its usefulness as a basis for payment, as a valid measure of efficiency in much narrower fields than it is often used. Thus the time series of Index Numbers of productivity for industries, *eg*, those published by *The Guardian*, may be useful in estimating the speed and direction of productivity changes. For reasons already given they are not measurements from which we can make inter-industry or international comparisons of efficiency levels. Almost any investment in machinery will increase labour productivity but only some investments will improve the efficiency of the firm – to which we now turn.

2 *The Firm*

The concept of efficiency applied to a firm relates to the input/output ratio. Usually both have to be converted into money values to enable a comparison to take place. Where output is a single product, a physical measure may be used, *eg*, the cost per ton of coal produced by the NCB, or by a particular pit. Considerable empirical work has been done in this field, both by firms who are anxious to check that they are producing at lowest cost per unit, and by economists seeking to assess actual cost patterns. The results of such inquiries should be examined in the light of the problems associated with this type of measurement, of which some examples are given below.

A major difficulty is to establish which costs should be included. Obviously this depends on the scope of the unit being examined, *eg*, a process, a plant – which may contain several processes – or a firm. If plant or process measures are the subject of the inquiry then the findings will probably concern only some costs or involve estimates or assumptions about head office charges, transport costs etc.

A further problem is that costs, in contrast to physical quantities of resources, are not only affected by the environment in which the firm competes to buy resources and sell its output but themselves affect the environment. Thus spending on advertising both depends on the environment and seeks to vary it by obtaining a market advantage. (See Chapter 5.)

For the economist, a reasonable level of profit is a part of cost of production, since it provides payment to the owners for the capital and enterprise which they have put into the firm. For the owner-manager, profits are the reward, firstly for his putting money capital into the business and secondly, for spending time in directing its operations. For the shareholder, only the first reason applies but in both cases the profits earned should be contrasted against the costs involved. These costs are both the alternative earnings of the money capital which have been forgone, *eg*, the current rate of interest, *and* the unavoidable risk that insufficient profits will be forthcoming to cover those costs. If such a level of profits is not forthcoming then resources will be directed elsewhere. Any profits above this level will be associated with competitive imperfections in the market mechanism (*eg*, monopoly, price agreements, and cartels) and are not costs of production. Monopoly power will enable a firm to buy inputs at below the competitive level. For the economist, the ratio of 'cost of all inputs'/'value of all outputs' will be 1/1 where the firm is in a very competitive situation. In a similar way the account's Profit and Loss Account always 'balances' (the difference being profit or loss). But whereas this is always so in accountancy, the 1/1 ratio to the economist would be upset by 'monopoly profits'.

So in a world of competition all ratios will be 1/1 using the economist's notion of reasonable profit as a cost. Different ratios can always be discovered if some costs (*eg*, the cost of capital which is omitted when measuring labour productivity) are excluded.

Comparisons of this type of (input-output) efficiency are possible inside an industry, between industries and between nations – so long as like is being compared with like. This will very rarely be the case since few firms sell the same range of products. To compare cost per unit of output between the Reliant Car Co, Rolls-Royce, and Ford would say little about their relative efficiency.

Fundamentally this is because although all three sell cars they are not competing in the same market and as a result will be producing at different scales of output. There is no smooth transition from one scale *eg*, Reliant's, to another, *eg*, Ford's because a large scale of operation would involve different manufacturing methods. Thus glass-fibre bodies are cheaper to produce in small quantities than pressed steel ones and so Reliant Cars have one form of body and Fords the other. This means that we cannot compare efficiency, only perhaps performance – and our performance comparisons have no more relevance between these three firms than between a butcher, a baker, and a candlestick-maker.

A further practical complication of any efficiency measurement is that different plants, built at different times, embody different states of technical knowledge. Again we can associate differences in the speed with which technical knowledge is used by a firm with performance; but only over a fairly long period of time and without measuring efficiency.

Another difficulty in making comparisons of efficiency using this measure is the well-known tendency for costs per unit of output to fall with the size of output (economies of large scale production) and rate of output (the learning effect). Thus the cost of a Ford Cortina will decrease as output increases because certain inputs are indivisible, *eg*, the cost of setting up body presses remains much the same whether 1 or 1 million are produced but this part of total cost per unit falls considerably with output. A jet aircraft becomes cheaper to produce the larger the production run of a particular type, due to the learning effect, *eg*, the skill of the labour force increases with repetition.

The meaning of our results, as always, depends on the reason for the inquiry. If we are merely looking for lowest costs per unit produced, then in many cases the prize will go to the largest plant, or firm, measured. If we are trying to establish comparative efficiency then we look for lowest costs at a given output. The distinction is important. For example, if we were considering a government policy favouring efficiency the former result would normally cause us to support the largest scale enterprise. The latter comparison may show up a well-managed small firm which at a smaller output is more efficient than the largest firm would be at that level and would continue to be so if allowed to grow to the larger scale output.

So this measure of efficiency which relates costs to output, like labour productivity, tends to ignore some costs, or assume they are constant. Their fundamental limitation for the manager is that *firms are usually run not to be efficient but to make profits.* An engineer may quite properly see, as his objective, the running of a plant at that output which gives the lowest average cost per unit, or at the lowest cost for a given output. These are very rarely the same thing. Most plants produce at lowest cost per unit of output when run at 'full capacity'. But trading conditions may require the plant to be run below capacity or at over-capacity – when the objective becomes lowest cost per unit for that level of output. Those who control the firm are likely to run the plant at the greatest profit – which is a measure of performance not efficiency.

In using maximum profit as the performance measure of a firm we are not claiming that it is the only possible test of performance. Others could be: survival;* the largest possible market share; sales (which may lead to expansion in new markets); size (measured in a variety of ways); rate of growth, and providing a good service to the community in terms of price and wages paid. Some of the non-profit objectives may be pursued in the mistaken belief that they

* The problems of dissimilar outputs etc, which make studies of comparative average costs difficult in most industries, have led to a technique whereby an industry is divided into ranges of size and changes over time in the proportion of industry output coming from a size class measured. The most successful size of a firm is that whose size class increases its market share by the greatest amount. This says nothing about absolute and only a little about relative efficiency; it does measure performance particularly against the goal of survival. The Lancashire textile industry has recently provided several examples of firms whose labour, and capital, productivity were rising but who made losses and were taken over.

are another way of maximizing profits. So a sales department may maximize sales even though some orders cost more in transport charges than they contribute to revenue. Few businessmen, or economists, believe that firms take all decisions rationally with profit maximization as the objective. Many managements probably look firstly for enough profit to keep the shareholders happy and then apply any surplus to growth – or some other objective.

Yet another set of objectives are appropriate to non-private enterprise organizations. Nationalized industries have already been mentioned in Chapter Four. An interesting case is that of local government authorities who are judged by their success in keeping rates low rather than by such criteria as maintaining efficient services etc. We assume that the justification for this is that to provide a given level of services at lowest cost is to be efficient. But surely this mis-states the objectives of a local authority which is overall to balance the desires of rate-payers for services with their willingness to pay and in particular to shift resources between services.

We are not saying that one objective is morally superior or socially more desirable than another nor have we the space to explore the ramifications of following the alternative goals suggested above. The field will now be narrowed to profits since most firms use this as a goal in one form or another.

Now it is obvious that the statement 'Profits this year £1 million' is meaningless in terms of performance. It must be related to capital employment, or sales, to enable this year's performance to be compared with last year's or our firm to be compared with another firm. There may be technical arguments among managers as to which relationship is most meaningful. Thus capital employed is usually preferred to sales or purchases or other possible measurements. Basically, the relevance of such a measurement depends upon the onlooker's viewpoint. For a financial investor or a member of the firm an increase in profits is clearly good but for the public as consumers there may be disadvantages in the form of higher prices associated with monopoly power.

Sometimes profits are judged by a further criterion – that of consistency with last year's results. So managers may take decisions with the object of achieving stable profits. In addition, firms may follow a policy of steady distributions of dividend even where profits fluctuate. This latter practice is unexceptionable so long as the profits information is available to the market. Despite special pleading, such as the bank's defence of their, until recently, secret reserves on the grounds that public confidence would be damaged if the true fluctuations were known, anything which distorts true profitability in the price mechanism runs contrary to the public's interest. One example would be a depreciation policy designed to promote a smooth flow of profits. We rely on the private sector to allocate resources between projects (and firms and industries) – how can this be effectively done when the true signals from the system are concealed?

Furthermore, the efficiency of the price mechanism does not require minimal fluctuations in the activities of a firm. We should expect some firms to do better than others in a competitive environment; dampening down fluctuations in the individual firm reduces the chances of gains from more efficient firms. Of course, fluctuations do involve costs which may present problems if the firm is large and, for example, the labour force removed by a down-turn in profits finds difficulties in obtaining jobs elsewhere. The implications for the economy of the inflationary pressure generated by high-earning car workers and the sudden reductions in purchasing power when, every few years, a proportion are laid off, is an example of this. Fluctuations in profits may be disliked by investors. In this case the solution is normally quite simple – for the investor to diversify his share-holding.

How is profit used as a measurement of performance? Economists, in their theories of how firms behave, usually measure profit per item related to price. Their theories are too general to give much guidance to assist such bodies as the Monopolies Commission, who are trying to assess what is 'reasonable' profit margin.

More important are the uses of profit made by managers and the general public in assessing performance. Management accountants have three widely used measurements:* *viz* Profits/Sales (P/S); Profits/Net Assets (P/NA) where 'net assets' are (net fixed assets + current assets) less current liabilities and provisions, and Profits/Total Assets (P/TA).

Profits are usually adjusted to allow for depreciation and may be pre- or post-tax. A whole battery of ratios is now available to the manager. Many do not use profit at all but provide partial indicators of performance in the firm and may, in addition, attempt to forecast financial problems. Examples would be: Sales/Total Assets; Debtors/Sales, and Stock turnover.

To use a selection of these ratios for information on which managerial decisions can be taken *within* a firm is a useful technique since a change in a ratio will form the basis of inquiries to find out why. Thus a Sales/TA ratio will worsen if a firm grows by taking over one of its suppliers since Sales may not change but Total Assets will increase. But the effect of the takeover on the ratio can be allowed for by management. Similarly this type of growth will improve a Profits/Sales ratio so long as the formerly independent supplier makes any contribution to group profits. But this highlights the limitations of such ratios in making inter-firm or inter-industry comparisons. They are of little value to an outsider hoping to measure efficiency or performance.

There are two major reasons for this. Firstly, any ratio which involves assets (TA or NA) runs into the accounting convention of measuring the value of assets by the price paid rather than their present market value. In the published accounts of firms it is normal practice to use this 'historical cost' valuation (adjusted by depreciation). This may vary

* A useful analogy with what follows for some readers may be with Standard Costing techniques. These are widely used methods of comparing performance (actual costs) with a standard, and great attention is given to the 'correct' setting of standards. But no accountant would dream of using them as a basis for inter-plant or inter-firm comparison.

widely from the present market price and differences may be great within the same industry. The difference between the two measures reflects changes in the value of money and changes in costs of production and demand for the assets. There is, anyway, a philosophical problem that the present value of an asset is related to the expected future flow of profits which it can be used to generate. Thus P/TA is a ratio, both parts of which are affected by present and future profits in so far as the assets are valued with reference to their earning power and not simply their historical cost.

Economists get upset about these differences because one of their major interests is the use of resources. Unreal valuations of the resources controlled by a firm are likely to lead to incorrect managerial decisions, not only in terms of performance related to an objective (profit maximization *et al*) but in terms of using resources where they can be most effective. So a firm which misleads itself into thinking that it is obtaining a reasonable return on capital employed only because it is valuing, say, freehold properties bought years ago at cost and not current market value, is not only fooling itself but is misleading the price mechanism. Using these unrealistic measures may enable it to continue to command resources which could be more effectively used elsewhere.

Secondly, different firms and industries vary in the convention which they use to estimate depreciation. This affects net profits as well as assets.

Having pointed out their shortcomings, we must also add that inter-firm comparisons can be possible between firms in the same industry if great care is taken in measurement and use is made of a variety of ratios. In some industries inter-firm comparisons are made using commonly agreed ratios. Each firm submits information in confidence and receives best, worst, and average results for the group. Considerable sophistication is possible in such schemes run by the Centre for Inter-firm Comparison.

But it is not as if something better than the crude measurements of efficiency does not exist. Professor Ball[1] has developed a measurement of efficiency, rather than performance, which relates inputs (in the form of capital and labour) not to output, but to 'Value Added' – (defined below). The information required is available inside a firm, although it is not normally generated since it is not required for published accounts. This has so far restricted inter-firm and inter-industry comparisons using the technique.

There are two great attractions in this approach. Firstly, we are measuring efficiency in terms of more than one factor. We have seen that (labour) productivity is a measurement in terms of only one factor; similarly ratios comparing profits or sales to assets or 'capital employed' measure 'capital productivity'. Professor Ball points out the anomaly of the nation being concerned with labour productivity and with managers being concerned with capital productivity. This implies that either labour or capital is the scarce resource. Labour is not the sole limiting factor on the nation's resources, nor is capital the only barrier to growth for a firm. Firms may set up their own barrier by decisions to limit borrowing. In fact, of course, both factors are scarce, since their total amount is limited at any one time and they are available in total only in a fixed proportion, although they will be used in different proportions by an industry depending on the process of production.

Secondly, a high proportion of a firm's costs of production consist of payments for raw materials etc bought in, where the price is not controlled by the purchaser. Value Added (Sales LESS raw materials, services, and sub-contracted work) is the measure of work done by a firm.* Value Added considered from the payments approach comprises gross profits plus remuneration to the workforce. If we are interested in efficient use of resources it is the relationship between work done and resource use which is important *NOT* sales – which are affected by the monopoly element in prices for outputs. Nevertheless, we have still not escaped completely from the influence of performance (as opposed to efficiency) since Value Added includes gross profits –

* Value Added of all firms totals Gross National Product. See Chapter Three.

which are usually assumed to be the firm's objective in terms of which efficiency is measured.

In the article referred to above, output is assumed to be homogeneous and only two inputs are measured. This allows the relationship to be shown on a graph. There are two limitations to the whole approach; it is necessary to assume constant or increasing returns to scale (*ie*, doubling Value Added doubles, or less than doubles, inputs) and costs such as advertising and research and development which change the economic environment must be excluded.

Even so, the insights following from the analysis are remarkable. It is possible to analyse total efficiency into a technical (physical input-output) and a price component (the extent to which a firm uses the least cost combination of capital and labour, *ie*, its success in reacting to relative factor prices). It is even possible to indicate the pace and direction of changes in efficiency. Rough inter-industry comparisons can be made of the efficiency of industries such as chemicals (with high labour productivity) and construction (with high capital productivity). At the time of writing further development of this approach continues in 'Business Ratios'.

We can now summarize much of what we have said about the difficulties of measuring efficiency with the following examples.

A firm might have the physical output/worker ratios shown in Table 25.

TABLE 25

Output	Nos of Workers	Output units per Worker
1000	10	100·0
2000	18	111·1
3000	24	125·0
4000	36	111·1
5000	50	100·0

a. Clearly, there is some sense in which the output of 3,000 is the most 'efficient' in that it has a more favourable labour/output ratio.

b. 'Performance', on the other hand, would compare these outputs with those of a second firm which might have the following output units per worker in Table 26.

TABLE 26

Output	Nos of Workers	Output units per Worker
1000	15	66·6
2000	20	100·0
3000	23	130·4
4000	40	100·0
5000	60	83·3

The 'performance' of the first firm is greater at all output levels except its most 'efficient' level of 3,000. The concepts of 'efficiency' are quite separate and independent; some secondary schools, for instance, have a higher 'performance' than others in terms of examination passes, but they may be less 'efficient' than others in terms of the real inputs of time and resources for each pass achieved. Parents are perfectly free to like a school because it has a high 'performance' for their children but they should not mistake this for efficiency.

Let us turn now to the relationship of money input and output as shown in Table 27.

TABLE 27

Output	Other Costs (£)	Nos of Men	Wage rate (£)	Wage bill (£)	Labour Cost per unit (£)
1000	1000	10	20	200	0·20
2000	1000	18	20	360	0·18
3000	1000	24	20	480	0·16
4000	1000	36	20	720	0·18
5000	1000	50	20	1000	0·20

a. In these circumstances, there is, of course, no merit to be attributed to the first 24 workers, simply because labour costs minimize at 24 workers. The same 24 workers are to be found in the totals of 36 and 50 workers.

b. There is no need to equate the 'money efficient' output level of 3,000 as consistent with the objectives of either the firm or society. The firm might be interested in profit maximization which could be at a quite different output according to the product prices. Society, in its turn, might prefer a zero output with all the resources used elsewhere.

c. If we lower the wage rate then the firm would have lower wage costs and an improved 'money efficiency'. But society would be neither better nor worse off; in real terms it forgoes the same resources to gain the same output. Similarly, a doubling of product prices would double the firm's money efficiency but would do nothing for society as a whole; individual members would, of course, be richer or poorer.

d. We have assumed that the wage rate does not have to be raised to attract more workers. If this should be the case, Table 28 might result.

TABLE 28

Output	Other Costs (£)	Nos of Men	Wage rate (£)	Wage bill (£)	Labour Cost per unit (£)
1000	1000	10	20	200	0·20
2000	1000	18	25	450	0·22
3000	1000	24	30	720	0·24
4000	1000	36	35	1260	0·31
5000	1000	50	40	2000	0·40

Suddenly, with no change in the real effectiveness of labour, the 1,000 unit output becomes the least labour cost level of production. Once again we see that there are few general rules in business policy; a change in the environment of the firm radically alters its position – in this case by changing the efficient output level.

e. Returning to our figures in Table 27 we can demonstrate that while 3,000 was the lowest money/cost output when other costs were given at all outputs – in this case, they happened to be £1,000 – this will change if these costs, too, alter with output. We must now look at total cost per unit in Table 29; in so doing we shift our focus from a concern with a single input to all inputs.

TABLE 29

Output	Other Costs (£)	Nos of Men	Wage rate (£)	Wage bill (£)	Labour Cost per unit (£)	Total Cost (£)	Cost per unit (£)
1000	1000	10	20	200	0·20	1200	1·20
2000	2000	18	20	360	0·18	2360	1·11
3000	2500	24	20	480	0·16	2980	0·99
4000	3000	36	20	720	0·18	3720	0·93
5000	3500	50	20	1000	0·20	4500	0·90

Both the lowest labour cost and lowest total cost in Table 27 were at an output of 3,000 units. In Table 29 lowest labour costs do not correspond with lowest total costs which now occur at the 5,000 output level. *Thus we cannot look for money efficiency by looking for laboour productivity.*

3 *The Industry*

Most of the problems of increasing efficiency, or performance, have already appeared in the previous section. This reflects the emphasis of economic analysis, while public discussion often concerns itself with the efficiency (or performance) of an industry, most recent work by economists takes the firm as the unit to be discussed. However, a little more remains to be said.

The whole industry may differ from the sum of its firms in the sense that, in the same way that size in a firm can bring the economies of large-scale production mentioned earlier, so growth in an industry may reduce the costs of inputs to the individual firm. This comes about through external economies. The classic example is in coal mining where the sinking of an additional pit in an area reduces the water level in other pits and thus lowers their costs of production. Another example is the lowered costs of training labour which follow from an industry being sufficiently concentrated in an area for specialized educational facilities to develop. In principle, external economies are another example of 'externalities', mentioned in Chapter One, which arise through the distinction between private and social costs and benefits.

The concept of the industry is most important if we attempt to consider an industry as an efficient user of resources. One important element in this would be external economies, but of greater importance would be the effect of the structure of the industry (*ie*, what difference does it make if a given car output is produced by 1, 10, or 100 firms). We are now considering the industry as a part of the market mechanism. Unfortunately we have only general predictions about the effects of structure on efficiency or performance.

The justification for the market mechanism rests on conditions of effective competition and in Chapter Five (page 122) we considered the problem of intervention with this as the objective. Market imperfections tend to have industries as their base. Thus, where it is possible, *eg*, by owning the world supply of a raw material, to make entry to the industry difficult or impossible, a portion of the environment can be made less competitive and thus possibly less efficient in terms of resource use.

4 *The Government*

We remain unconvinced that there are any special reasons for treating government efficiency differently from the efficiency of a firm. Even the problem of there being no adequate basis for a comparison can be considered as merely an example of the general difficulty of inter-industry comparisons. Anyway, in many areas of government activity it is fruitless to compare the government's efficiency or performance with a hypothetical private firm since the private sector just cannot do the job (*eg*, defence, regional policy).

Turning to efficiency, there is no evidence that, as a user of resources, the government is less efficient than the private sector. This applies to the government as a consumer *eg*, National Health Service, investor, *eg*, road-building programmes, and as a producer, *eg*, the nationalized industries. Nor is there any evidence in the opposite direction (the necessary research has just not been done), except perhaps a presumption that with the recent development and use of sophisticated investment techniques (and, *eg*, cost-benefit analysis) the public sector is probably more advanced than all but the best of private enterprise. It is not very clear how such comparisons could be made since, *eg*, nationalized industries have different objectives. Given the different objectives in the public sector and given that there is no evidence that private enterprise is absolutely efficient it would seem fruitless to make comparisons of relative efficiency between them. Governments can be judged in terms of their own objectives, enough has been said to point out the inapplicability of private sector objectives.

Private citizens have a special attitude to government expenditure based on an interest in what 'they' are doing with 'my' money. This unease is not dispelled by the inadequate parliamentary controls over expenditure, particularly due to the way in which expenditure is presented for approval. Very rarely can the observer compare the cost of a programme with the benefits to be received. An exception is defence programmes, which are sometimes presented in the form of a comparison of levels of spending and military equipment, establishments etc, to be maintained. But the implications of health, transport, education programmes etc, are rarely described in this way with all costs, current and capital, associated with all benefits. Such a presentation is the first step towards control. The second is to compare the advantages to be gained by spending (or withdrawing) the last £1 million from one programme and transferring it to another. Assuming that the benefits could be measured, this provides a technique which would maximize the value to be gained by society from the marginal expenditure. The analogy with the operation of a firm is obvious.

Basically, we judge a government by its success in attaining objectives.* In the past these have been financial. Thus in the 19th century the aim was to refrain from intervention whenever possible and to keep costs, and thus taxation, to a minimum. It was believed that financial prudency required that income and expenditure for a year should be balanced. This emphasis on avoiding 'costly' intervention allowed, and probably favoured, tariffs, a form of intervention which although cheap in terms of government expenditure imposed widespread social costs and benefits. By the end of the First World War the overriding emphasis on minimal taxation had been relaxed. The adoption of Keynesian analysis in UK Budgets since 1941 has meant that this objective can be rephrased as setting taxation and expenditure, not to be identical in accounting terms but to produce a balanced economy in the Keynesian sense.

We are left with judging a government not by efficiency but by performance. The objectives could be those specified earlier in Chapter One but, as was pointed out then, readers can add to and remove items from this list. Whereas the public may think that they have a criterion for efficiency in a firm, we are quite clear that our different prejudices result in there being no agreed list of objectives ranked in order of importance. Thus there are no agreed criteria of importance.

5 *The Economy*

While no measurement is possible, it can be convenient to think of an economy's *efficiency* as the relationship between GNP and the national stock of factors of production. In this sense higher 'productivity' of any factor – *land*, *labour*, *or capital* – can be seen to be in the national interest. Thus, if a firm uses only one input in a process, an increase in productivity is a benefit to the firm and indirectly and imperfectly, since we do not have perfect competition, to society. But firms very rarely use only one factor and if they

* A limitation of the authors, and of the scope of this book, is that we are solely concerned here with economic objectives, whereas a government is judged in much wider political terms.

ignore others then either their efficiency measurement misleads them or it misleads society. The scope for all firms to change performance, however, is clearly limited by the proportions in which factors are available to the economy as a whole. The only way in which this stock of resources can be changed is through investment, in its broadest sense, which will affect the quantity, quality, and relative importance of resources.

National average figures for the efficiency of one factor only will not do since technology determines that some industries will have, say, higher labour productivity, than others. The national figure may rise simply because of a growth in the relative importance of such industries in the economy, even though productivity did not increase in a single firm.

The importance placed by our society in the market mechanism as a device for allocating resources is now clear. So long as we use quantitative measurements of efficiency we require profit maximization as an objective in order to maximize outputs from our inputs.

A consideration of the *performance* of the society in which we function must be very largely a value judgement, based on social and political matters, so that as economists we withdraw just when the discussion is becoming interesting to leave the field to political philosophers, sociologists *et al.* In economic terms the government's objectives must presumably be at least those held by a majority of the electorate at any date up to five years ago.* But, in a wider context, how important are they compared with the quality of like terms of freedom, social justice, amenities etc? An excellent

* This is meant to be ironic rather than naïve. Clearly governments are influenced by pressure groups during their term of office. In addition the multiplicity of objectives (see the list of merely economic aims page 252) means that, conceivably, a major objective in the electoral manifesto of a successful Party may be disliked by a majority of its supporters – and perhaps by all the opposition. The possibilities are endless, but do point to the futility of using a referendum on such questions as, *eg*, should we increase unemployment to have a better Balance of Payments?

case,[2] in economic terms, has been put forward to show that although increases in GNP may measure the overall economic growth this may cause a worsening of the non-priced items in our environment (*eg*, quietness) and may adversely affect some groups.

Conclusion

Having got this far, the reader may feel that, even more than some others, this has been a negative, unconstructive chapter. Certainly the inadequacies and difficulties of efficiency measurements for the individual and the firm have been dealt with at length. In addition we have been concerned to point out the boundaries of the area which can be subjected to scientific analysis. All managers are concerned in their working activities with some aspects of efficiency; this chapter has attempted to put those partial measurements, which may be effective in a limited sphere, into a larger perspective and to point out some of the contradictions which are involved when only one factor of production is assumed to be important.

However, we feel that other parts of this chapter can be summarized in a more optimistic way. Productivity agreements have potential benefits which have not been stressed so far, since our emphasis has been on the problems and complexities of measurement. Such agreements may affect other measurements of efficiency, as in cases where overtime is reduced and shift work is made possible – which will improve the capital/output ratio. Other productivity agreements cause shifts to more capital-intensive or labour-intensive methods of production. The latter would follow from greater output associated with a fixed stock of equipment. If the former effect occurs, due perhaps to increased labour effort making the installation of capital equipment worth while, the resulting increased investment may be seen as desirable from the point of view of growth. To present this as a costless form of growth is misleading. Whereas one firm may increase its investment, it can only do this at the expense of another so far as a given stock of capital is concerned; if however, and it seems more likely, the advantage is expected to come from a greater rate of overall investment, then the costs involved are the lost consumption which must be forgone in order to allow the increased production from the capital goods industries.

Labour productivity remains as a valuable concept applied within the individual firm, and there is a growing awareness of the importance of the twin concept of capital productivity. This is associated with a recognition that, in addition to land and capital, firms possess a third resource – labour – and that the notion of efficiency measurement can be applied to all three.

We can hope for a more widespread appreciation of the efficiency of a firm. Partly this comes from greater disclosure of the relevant information to the rest of the market system. Partly firms are becoming better informed about their efficiency and performance, as management accountants produce more and more information which is not simply a reorganization of the historical records of the business but is based on valuations and analyses which are relevant to decision taking by managements. Concerning the efficiency of the economy, a start will have been made when the problems caused by the multiplicity and subjective nature of the objectives are recognized.

Great importance is attached in society to efficiency and performance; great precision and care is necessary in measuring them. In the end 'economic performance can only be assessed against aims, needs, and potential'.[3]

References

1 Ball, R. J., *op. cit.*

2 Mishan, E. J., *The Costs of Economic Growth*, Staples, 1967. Also by Penguin Books.

3 Lomax, K. S., *op. cit.*, page 9.

Bibliography

Amey, L. R., *The Efficiency of Business Enterprises*, Allen and Unwin, 1969.

Ball, R. J., 'The Use of Value Added in Measuring Managerial Efficiency', *Business Ratios*, Summer 1968.

Dunning, J. H., 'US Subsidiaries in Britain and their UK Competitors', *Business Ratios*, Autumn 1966.

Lomax, K. S., *The Assessment of Economic Performance*, Leeds University Press, 1965.

Speight, H., *Economics and Industrial Efficiency*, Macmillan, 2nd Edition 1967.

Index

MANAGEMENT SERIES

MANAGEMENT DECISION MAKING (30p)6/-
A symposium of five international experts—British and American—stress the importance of scientific decision making in modern business administration.

MARKETING MANAGEMENT IN ACTION (60p)12/–
Victor P. Buell. A guide to successful marketing management by a former national vice-president of the American Marketing Association.

THE PRACTICE OF MANAGEMENT (50p) 10/–
Peter F. Drucker. An outstanding contribution to management theory and practice.

MANAGING FOR RESULTS (40p) 8/-
Peter F. Drucker. A 'what to do' book for the top echelons of management.

THE EFFECTIVE EXECUTIVE (35p) 7/-
Peter F. Drucker. How to develop the five talents essential to effectiveness and mould them into results by practical decision-making.

CYBERNETICS IN MANAGEMENT (40p) 8/-
F. H. George. Introduction to the ideas and methods used by cyberneticians in the running of modern business and government.

PLANNED MARKETING (30p) 6/-
Ralph Glasser. A lucid introduction to mid-Atlantic marketing techniques.

FINANCE AND ACCOUNTS FOR MANAGERS (30p) 6/-
Desmond Goch. A vital and comprehensive guide to the understanding of financial problems in business.

INNOVATION IN MARKETING (37½p) 7/6
Theodore Levitt. A brilliant exposition of original and stimulating ideas on modern approaches to marketing.

MANAGEMENT SERIES (cont.)

THE ESSENCE OF PRODUCTION (40p) 8/-
P. H. Lowe. Explains the components, diversities and problems of production within the general framework of business management.

MAKING MANPOWER EFFECTIVE (Part 1) (37½p) 7/6
James J. Lynch. The techniques of company manpower planning and forecasting.

SELLING AND SALESMANSHIP (25p) 5/-
R. G. Magnus-Hannaford. A clear, concise and forward looking exposition of practical principles and their application.

CAREERS IN MARKETING (30p) 6/-
An Institute of Marketing Review. A guide to those seeking a job in the exciting field of marketing.

THE PROPERTY BOOM (illus.) (37½p) 7/6
Oliver Marriott. The story of the personalities and the companies that emerged enriched from the commercial property industry in the years 1945-1965.

MARKETING (37½p) 7/6
Colin McIver. Includes chapters by Gordon Wilson on The Years of Revolution and Industrial Marketing.

EXPORTING: A Basic Guide to Selling Abroad (37½p) 7/6
Robin Neillands and Henry Deschampsneufs. Shows how smaller and medium-sized companies can effectively obtain and develop overseas markets.

DYNAMIC BUSINESS MANAGEMENT (25p) 5/-
Harold Norcross. A simple guide to the rudiments of successful business management.

FINANCIAL PLANNING AND CONTROL (40p) 8/-
R. E. Palmer and A. H. Taylor. Explains the nature of the assistance which higher levels of accounting can provide in the planning and control of a modern business.

MANAGEMENT SERIES (cont.)

COMPUTERS FOR MANAGEMENT (30p) 6/-
Peter C. Sanderson. A timely appraisal of computers and electronic data processing—their basic concepts, potential and business application.

GUIDE TO SAMPLING (30p) 6/-
Morris James Slonim. A fine exposition of sampling theory and techniques.

MANAGEMENT INFORMATION—Its Computation and Communication (40p) 8/-
C. W. Smith, G. P. Mead, C. T. Wicks and G. A. Yewdall. Discusses Education in Business Management, Statistics for Business, Mathematics and Computing, Operational Research, Communicating Numerical Data.

THE REALITY OF MANAGEMENT (35p) 7/-
Rosemary Stewart. Compass bearings to help the manager plot his career.

MANAGERS AND THEIR JOBS (35p) 7/-
Rosemary Stewart. Helps managers to analyse what they can do, why they do it, and whether they can, in fact, do it better.

These Management Series titles are obtainable from all booksellers and newsagents. If you have any difficulty please send purchase price plus 9d. postage to Claude Gill Books, 481 Oxford Street, London, W.1. where the whole series is on display.
While every effort is made to keep prices low, it is sometimes necessary to increase prices at short notice. PAN Books reserve the right to show new retail prices on covers which may differ from those previously advertised in the text or elsewhere.